Network Centric Warfare and Coalition Operations

This book argues that Network Centric Warfare (NCW) influences how developed militaries operate in the same fashion that an operating system influences the development of computer software.

It examines three inter-related issues: the overwhelming military power of the United States; the growing influence of NCW on military thinking; and the centrality of coalition operations in modern military endeavours. Irrespective of terrorist threats and local insurgencies, the present international structure is remarkably stable – none of the major powers seeks to alter the system from its present liberal character, as demonstrated by the lack of a military response to US military primacy. This primacy privileges the American military doctrine and thus the importance of NCW, which promises a future of rapid, precise, and highly efficient operations, but also a future predicated on the 'digitisation' of the battlespace. Participation in future American-led military endeavours will require coalition partners to be networked: 'interoperability' will therefore be a key consideration of a partner's strategic worth.

Network Centric Warfare and Coalition Operations will be of great interest to students of strategic studies, international security, US foreign policy, and international relations in general.

Paul T. Mitchell is Associate Professor with the Department of Defence Studies, Canadian Forces College in Toronto.

Routledge global security studies

Series editors: Aaron Karp, Regina Karp and Terry Teriff

1 **Nuclear Proliferation and International Security**
Sverre Lodgaard and Morten Bremer Maerli

2 **Global Insurgency and the Future of Armed Conflict**
Debating fourth-generation warfare
Terry Terriff, Aaron Karp and Regina Karp

3 **Terrorism and Weapons of Mass Destruction**
Responding to the challenge
Edited by Ian Bellany

4 **Globalization and WMD Proliferation**
Edited by James A. Russell and Jim J. Wirtz

5 **Power Shifts, Strategy and War**
Declining states and international conflict
Dong Sun Lee

6 **Energy Security and Global Politics**
The militarization of resource management
Edited by Daniel Moran and James A. Russell

7 **US Nuclear Weapons Policy After the Cold War**
Russians, 'rogues' and domestic division
Nick Ritchie

8 **Security and Post-Conflict Reconstruction**
Dealing with fighters in the aftermath of war
Edited by Robert Muggah

9 **Network Centric Warfare and Coalition Operations**
The new military operating system
Paul T. Mitchell

Network Centric Warfare and Coalition Operations

The new military operating system

Paul T. Mitchell

Routledge
Taylor & Francis Group
LONDON AND NEW YORK

First published 2009
by Routledge
2 Park Square, Milton Park, Abingdon, Oxon OX14 4RN

Simultaneously published in the USA and Canada
by Routledge
52 Vanderbilt Avenue, New York, NY 10017

First issued in paperback 2013

Routledge is an imprint of the Taylor & Francis Group, an informa business

Typeset in Times by Wearset Ltd, Boldon, Tyne and Wear

British Library Cataloguing in Publication Data
A catalogue record for this book is available from the British Library

Library of Congress Cataloging in Publication Data
A catalog record for this book has been requested

ISBN13: 978-0-415-84904-3 (pbk)
ISBN13: 978-0-415-44645-7 (hbk)

Contents

Illustrations

Figures

Table

Acknowledgements

This book would not have been possible without the support and encouragement of many people. First, thanks must go to Dr John Cowan, former principal of Royal Military College for granting me an extended sabbatical from my administrative position at the Canadian Forces College. Second, my heartfelt gratitude goes to the staff and students at the S. Rajaratnam School for International Studies at Nanyang Technological University in Singapore, and in particular, to the Dean, Barry Desker in creating such a superb opportunity for me to teach and research in that wonderful country. I am also very grateful to Dr Amitav Acharya for his encouragement and assistance with this project. Finally, I am most grateful to the staff and students of my home institution, the Canadian Forces College in Toronto. I am frequently asked by my colleagues in other academic institutions if I would not rather work in a 'real university'. While the administrative burden of working within a staff college environment is frequently onerous, there are no better research opportunities in the defence and security area anywhere else in Canada. Indeed, this book is ultimately the result of a question posed by a former student in the College's erstwhile Advanced Military Studies Course. Concerned by the rise of security restrictions frustrating information sharing, LCol. Bob Chekan mused on the possibility that digital networks would ultimately result in 'clueless coalitions'. My hope is that this work will help to further the growing reputation of CFC as a credible centre of excellence in this discipline.

Many others have assisted in the production of this work. Greg Gilbert and David Stevens of the Royal Australian Navy's Sea Power Centre were instrumental in helping me reach all of the commanders of Australia's naval contributions to the War on Terror between 2001 and 2003. I am especially grateful to all those military officers, serving and retired, in Australia, Canada, the United Kingdom, and the United States who consented to participate in this project and for being so open and frank about the nature of their work in the field. Charles Pentland of Queen's University provided 'rapid and decisive' assistance in tracking down permission for obscure operational maps. LGen. Charlie Bouchard graciously permitted me to quote from his presentation to the students of the National Security Studies Program in Washington DC.

Parts of this book have been read by Guy Ben Ari, LCol. Derek Basinger, LCdr. Rick Booth, LGen. Charlie Bouchard, Adam Chapnick, Christopher Coker, Urmilla

Deshpande, Mary Dub, Deborah Elms, LCol. Barry Green, Emily O. Goldman, MGen. Fraser Holman (Ret.d), Tim Huxley, Col. H. R. McMaster, Joseph Jockel, Bernard Loo, Meithili Mitchell, Pierre Pahlavi, David Schmidtchen, and Joel Sokolsky. I am grateful for their efforts and keen observations on the chapters they assisted with. Any errors or omissions, of course, remain my full responsibility. Taylor & Francis Books for permission to reprint elements of 'Network Centric Warfare: Coalition Operations in the Age of US Military Primacy', Adelphi Paper 385, (London: Adelphi Papers, IISS, 2006), in Chapters 1–4. Chapters 1, 2, and 4 have been reworked to take account of shifts in research, and Chapter 3 appears unchanged. McGill Queens University Press to reprint Figure 5.1 'NORAD operational boundaries, circa 1961' originally published in Joseph T. Jockel's *Canada in NORAD 1957–2005: A History*.

Special thanks go to Geoffrey Till for assisting with my studies at the Joint Services Command and Staff College in Watchfield UK, and in particular, putting me up in his fourteenth-century cottage for three wonderful spring days in 'England's green and pleasant land'.

Introduction

> Everything has changed, except our way of thinking.
>
> Albert Einstein

> The change from atoms to bits is irreversible and unstoppable.
>
> Nicholas Negroponte

The world is seized with the idea that we are at the doorstep of a new society. While Einstein wrote of the implications of a nuclear age, our thoughts today are guided by the vision of a future enabled by the power of digital technology. Negroponte's observation above notes the inevitability of this shift, as have many others. Larry Ellison of Oracle has noted that all forms of knowledge will ultimately reside on the Internet: 'It's collecting all the knowledge of mankind and making it available in digital fashion – reliably, securely and economically.'[1] Howard Rheingold observed that the Internet is evolving into an 'innovations commons and laboratory for collaboratively creating new technology'.[2] Don Tapscott and Anthony Williams make the case for a whole new approach to business and economics stemming from the technological changes taking place in early twenty-first-century society.

> Billions of connected individuals can now actively participate in innovation, wealth creation, and social development in ways we once only dreamed of. And when these masses of people collaborate they collectively can advance the arts, culture, science, education, government and the economy in surprising but ultimately profitable ways.[3]

The implications of these shifts for military forces are only beginning to be grasped, but there is much to suggest that despite the ravenous appetite for information technology (IT), all is not well with how militaries will adapt to this new world. In truth, the same technology that is building new communities and enhancing people's ability to cooperate, collaborate, and communicate on a global basis may perversely limit military interoperability and thus the prospect for more multilateral and cooperative international ventures aimed at restoring and enhancing international stability, protecting populations under threat, as well as traditional military ventures between alliances and coalition partners.

These enormous changes are expected to have significant impacts on our society. Vincent Moscoe, in a survey of the literature that has accompanied the development of the Internet, has observed that these typically fall into three related themes – the end of history, the end of geography, and the end of politics. History will come to an end both in the sense that all that has gone before will be irrelevant in this new age, but also in the sense that new forms of community will be possible in an age defined by linkages between people. The importance of geography will diminish as new communication linkages will enable people to carry out their lives no matter where they are located, themes that have been taken up by journalists such as Frances Cairncross in *The Death of Distance*[4] and Thomas Friedman in *The World is Flat*.[5] Finally, the end of politics is presaged by a new liberty for the powerless to direct their destinies in radically new ways, enabled by the ability to bypass the traditional sources of power in the large monolithic institutions of human society – the state, the family, religion, and corporations.

While Moscoe points out that all these predictions have been made before in relation to other older forms of technology such as electrical grids, telegraph, radio and television,[6] Manuel Castells notes that there seems to be a distinct difference in the case of information technology associated with the Internet. Indeed, he claims that rather than the end of history, we may in fact be seeing the emergence of a new age:

> History is just beginning, if by history we understand the moment when, after millenniums of prehistoric battle with nature, first to survive, then to conquer it, our species has reached the level of knowledge and social organization that will allow us to live in a predominantly social world.[7]

Indeed, even stripping away the superlatives that often accompany much of the information age literature, there are enough changes manifestly evident to even casual users of IT to suggest that 'something is afoot', even if it is not the end of history.

Military interest in information technology predates that of contemporary society's. Indeed, the military can lay claim for developing much of the foundation for these technologies. Modern computers emerged during the heights of the Second World War in systems designed to assist with operations research in the fields of anti-submarine warfare and ballistics. Later, more famously, the development of the Internet itself was spawned by concern over the survivability of communications links carrying commands and orders to military forces fighting in the midst of a nuclear war. While the transistor and microchip were invented in commercial labs, the military invested heavily in them and arguably speeded up their development, power, and widespread introduction into civil society.

Today, commercial interest in information technology ensures that civil technology is generally more advanced than military applications. Indeed, in developing modern communication systems, civilian technologies, such as cellphones, often lead military systems by several generations in their applications and ease of use. This often leads to developmental problems in the form of military users demanding at least the same level of performance that they can get using the commercially available systems they employ in their everyday life. Besides this technological gap, there is also a gap within the literature analysing how these enormous social shifts in our culture will impact the military. While there are a significant number of publications that address the technological aspects of IT in the military, few consider the broader social implications that it will have on the military's role in the future.[8]

This is all the more surprising given the radical claims that accompany much of the 'Information Age' literature. Each of the three categories of change advanced by Moscoe plays an overwhelming role in the use of military force. Power, and thus politics are, of course, at the heart of the use of force in any age. Geography is central to the fighting techniques of military forces. Divided into their respective air, land, and sea branches, it is difficult to imagine the conduct of war fundamentally abstracted from the impact of geography. History plays such a dominant role in understanding the enduring aspects of warfare that modern commanders continue to study the battles of the ancients for understanding the challenges of the present. The curriculum of many war colleges, the Naval War College of the United States no less, use ancient texts such as Thucydides' *Peloponnesian War* as a foundational text in the study of the relationships between war and politics. If Castells is correct that we are at the beginning of a new age, then greater attention must be paid to these shifts.

These shifts, it is argued, are independent of any particular state or even region. While many have acknowledged the existence of a digital divide between those on the Web and those with no access, others have pointed out that such divides do not obey traditional geographic categories of north and south, centre and periphery. The digitally dispossessed can easily be found in major urban centres of modern Western cities, and cellphone networks are transforming societies in Africa and Asia. It is the globalised phenomenon of technological transformation that is perhaps the most remarkable aspect of this phenomenon. Still, where most military observers have focused their speculation lies largely at the state level of analysis.[9] Speculation has largely been restricted to the impact of high technology on militaries as monolithic entities, rather than internationally cooperative ones. Since the end of the Cold War, most developed militaries have been more active in internationally cooperative ventures such as peacekeeping and humanitarian operations as well as more forceful 'coalition operations'. In this new century, many militaries seem to have arrived at a strange period wherein they are as concerned with operating effectively with other partners as they are with mutual competition.

Furthermore, the present era of human history, whether it is on the verge of a new age or, in the inimitable words of Colin Gray, is simply *Another Bloody Century*, seems poised to demand increasing amounts of cooperation between military forces on a global scale. The challenges that are likely to confront all states in a period of global warming with all the transnational challenges such a shift portends, in a period where many of the global institutions, all developed in previous periods for different aims, come under challenge from globalised forces not fully under their control, and in a period in which, despite advances in communication technology and the advance of liberty, the concepts under which we govern ourselves seem increasingly contested. For all these reasons, as an executive arm of government, militaries will be important tools for delivering any proposed solutions or reactions to these globalised problems, and as globalised problems, cooperation and collaboration will be critical to implementation.

The technology that enables this globalised explosion of information sharing and digital collaboration, paradoxically, will inhibit the same in the military environment. Every day, new platforms and applications appear to assist individuals across the globe link up and form new communities online; on the other hand, militaries seem stalled in achieving similar levels of interaction. Network centric warfare (NCW) would seem to be the military analogue of civilian collaborative IT, and it is true that collaborative networks have been growing slowly within the boundaries of many militaries. However, international networks between militaries are far more rudimentary. Nor does this seem to be simply a factor of the lag between civil and military use of technology discussed above. Indeed, in many regards, the limited advances that collaborative digital technology has permitted in terms of networking between international groups of military forces may have already been fully realised, save only in the most extraordinary of circumstances. The limited forms of interaction already evident between coalition and alliance partners may be as good as it gets.

In this regard, even if we accept the contentious point that technology is eliminating the role of history, geography, and politics for civil society, each of these factors remains firmly entrenched in the military sphere. While technology may be flattening hierarchies on a global basis, on the battlefield such hierarchies are critical to survival, even on a 'network centric' battlefield.[10] The current military predominance of the US may in fact be reflexively reinforced by the shift to networking technologies, establishing an enduring hierarchy of the US and those actors it is willing to digitally cooperate with.

This study first examines the themes that emerge from the literature examining the information age. It notes that it is animated by an inherent anarchical spirit that mitigates against the security culture informing the military environment. The clash between the digital anarchism of the Internet and the hierarchical order of military, intelligence, and security organisations may frustrate their ability to replicate the online collaboration and creativity of contemporary society.

Next, it argues that irrespective of America's 'actual' power, its military primacy means essentially that it is the only state capable of acting anywhere on the face of the planet: all others are limited to their own strategic neighbourhoods, or capable only of short bursts of global activity. Use of American military power is increasingly complicated both by this fact and globalisation. Despite the fact that all major powers are currently in favour of the present international status quo, the 'risks' presented by the opportunities/challenges of globalisation will complicate international cooperation. In many cases, the US will stand alone in terms of decisions to intervene or not in the international sphere.

The theory of NCW will reinforce this unilateral drift. While the application of computer networks to military operations is sought primarily to increase the amount of operational freedom a commander has at his disposal, the twin needs to guard the security of a network's information and assure that it accurately represents current situational reality will mean that strict control of the information circulating on military networks will be necessary. This will raise significant issues for the ability of coalitions to share information amongst their members; network information assurance must come at the expense of coalition information release policy.

The technological limitations of coalition networks will themselves be reinforced by the political nature of coalitions and their management. Coalitions are largely about scarcity, either in terms of actual resources or political legitimacy. Scarcity is relieved through sharing influence over policy. The willingness to share influence is a function of how dependent a leader is on his followers. In the age of American military primacy then, influence will be tightly restricted to the very few partners who are capable, willing, and trusted to make meaningful contributions to US operations.

The study conducts an examination of how networks are affecting military operations. First, naval networks are examined through the case of coalition operations conducted in the Persian Gulf during 2002 and 2003 as led by America's Australian and Canadian coalition partners. This is a particularly critical case study for how NCW affects coalitions for a number of reasons. The Canadian and Australian navies are roughly similar in size, technical capability, and professionalism. They each share very similar and strong professional relations with the USN that extend back to the Second World War. Each played a leadership role for the coalition within their operational areas of the Persian Gulf. However, Canada and Australia pursued very different strategic policies with regards to the issue of Iraq in 2003. As such, the case study reveals the impact that strategic policy has over information sharing within coalitions.

Next, the potential of air-oriented networks is considered through an examination of the issues affecting NORAD. The ongoing development of missile defence centres in Europe has led to calls for a 'NORAD-East' by some. However, the history of NORAD suggests that this will not be a simple matter. The case is particularly of interest as it demonstrates the limitations that high degrees of professional trust between two military services play in enabling close strategic relationships between nations. While both the Canadian and American air forces enjoy a professional relationship similar to that of each nation's navy, NORAD has been buffeted by political forces of the Canada–US relationship. Paradoxically, even as the economic and civil infrastructures of the two nations grow ever more networked, the level of cooperation between Canada and the US in NORAD has been shrinking. The complexity of the post-9/11 security environment in North America together with the strictures of information security suggest that the objective of a seamless network architecture is unlikely to happen. This will be even more true of the missile defence centres springing up in

Eastern and Central Europe, where political and professional trust between America and its partners is even less well developed than that with Canada.

Finally, the study examines the significant problems that are confronting the application of NCW theory on land. Here, the land environment has a considerable impact on operations that raises real questions as to whether land-based networks will ever be as efficient as those deployed at sea or to monitor air operations. Taken in consideration of the coalition environment, the difficulty in sharing positional data of land units raises real questions on the future viability of coalition land operations. As land networks enable greater dispersion on the battlefield, the room for non-integrated military forces will progressively shrink, possibly to the point where forces can no longer be geographically separated within the battlespace. The three cases suggest that rather than assisting collaboration between military organisations, digital technology may have the reverse affect, stimulating unilateralism.

The information age

As Castells points out, as important as technology is, especially in the military sphere, it is only one part of the factors that go into the establishment of any particular society. Other economic, political, and cultural factors are all critical, first to the development of any particular technology, and second, to how effectively that technology is used and propagates within that society. Despite inventing critical technologies such as the compass, paper and moving type printing, and gun powder, China turned its back on all of these developments to such a degree that it was easily overwhelmed by Western powers in the eighteenth and nineteenth centuries who all employed such technologies to far greater effect globally.[11] For Castells, the shifts in global capitalist processes, the rise of social movements like feminism and environmentalism, and the origins of IT within the culture of American academe and then the liberal society of California are all as important a development in the shift to what he has called the 'Network Society' as the emergence of the Internet itself.[12] Indeed, the Internet would not be what it is today without these developments: some societies, notably the Soviet Union, were unable to exploit such technology to the same degree as the United States and other Western nations.

Individuals across the planet are taking advantage of the opportunities provided by such technology so that a new social form, which Castells calls 'informationalism', is replacing the structures, processes, and norms of industrialism. According to Castells, informationalism provides the foundation to this new network society 'based on the augmentation of the human capacity in information processing around the twin revolutions in micro-electronics and genetic engineering'. This network society has three essential features which distinguish it from how industrial and agricultural societies used information. First, is the ever expanding information-processing capacity of computer systems in terms of volume, complexity, and speed; factors all consistent with the so-called 'Moore's Law' of the processing power of semi-conductors. Second, is the ability of digitised information to be recombined with itself and other information endlessly, permitting a similarly reflexive development of innovation and creativity. Last, is the flexibility embodied in the nature of networks, permitting the widespread and uncontrolled spread of information generated by the first two factors.[13]

Other analysts have arrived at similar conclusions. Nico Stehr has described what he calls the 'knowledge society', which is distinguished by the increasing opportunity to act on the part of its constituents.[14] Valovic notes of its underlying ideology:

> The hope is that the Internet will forge in the white heat of information long kept compartmentalized, a new compact that a new view of the world will emerge from the dynamic of human history itself. In forging this compact, the Net's capacity to unleash the synergies of human thought long kept in abeyance by the entropy of institutions is paramount.[15]

Many analysts are careful not to whitewash the results of this development. Both Castells and Stehr are careful not to identify this emerging social form with any historical teleology or sense of human progress. Stehr notes that the knowledge transmitted under these conditions may be as contested as knowledge in any previous social form, and Castells notes that 'we do not really know if producing more or more efficiently embodies superior value in terms of humanity'.[16] Valovic agrees that the abundance of information enabled by such technology might eventually undermine all pre-existing standards on which to judge the good, the noteworthy, and the historical in return for the ephemeral.[17] Further, sociologist Scott Lash argues that the IT backbone naturally produces information overload,[18] and even Negroponte argues that digital multimedia lacks the richness of analogue literature because so little is left to the imagination.[19]

Nevertheless, as 'amplifiers and extensions of the human mind', new technology enables the creative and innovative powers of human thought as no other technology has in the past.[20] The unleashing of innovation and creativity has enabled the development of new power loci outside of traditional social structures. Thus, industrial titans are humbled by new start-up companies and states are stymied by amorphous social movements, just as militaries are challenged by insurgent movements.

Still, David Weinberger notes that this process is more about the links between information than its transformation into bits, or the much discussed empowerment of new groups. These 'loosely joined pieces' mark a fundamental movement away from the machine model of the world emerging from the age of enlightenment.[21] Castells remarks that this historical shift subverts the concepts of sovereignty and self-sufficiency that have guided the construction of identity since first discussed by classical Greek philosophy.[22] As the body becomes more and more plugged into the rest of the world, the boundaries between self and other become less and less distinct.[23]

This marks the current development of the Internet as much as a platform for computation and storage rather than simply a communicative medium. The emergence of peer-to-peer networks and distributive computing is resulting in the movement of data and applications away from the desktop computer onto the Web itself. The increasing interactivity of web sites on the Internet, referred to as Web 2.0, marks the difference from early web sites were users simply consumed the information posted, to active participation and community building applications like social networking sites.[24] Rheingold has described this shift between early forms of the Web to the present era in terms of the nature of the content on a web site. In early forms of the Web, 'content is king' and readers simply consume material offered. However, Web 2.0 sites encourage group-forming behaviour by enabling human communication to modify the content contained on the web site, 'jointly constructing value'.[25] Thus, the Web becomes a real collaborative space where content is created and developed by the users themselves. In this environment, it is suggested, new media companies such as Google, Yahoo, and YouTube unburdened as they are by historical legacies,[26] all have advantages over traditional media companies. As Castells concludes

> Networks are appropriate instruments for a capitalist economy based on innovation, globalization, and decentralized concentration; for work, workers and firms based on flexibility and adaptability, for a culture of endless deconstruction and reconstruction; for a polity geared towards the instant processing of new values and public moods; and for a social organization aiming at the suppression of space and the annihilation of time.[27]

Many of these themes are important for globalised and integrated military operations such as those characterised by the War on Terror. However, as will become apparent below, there are social impediments which will place frustrating barriers in the military's attempt to appropriate these technologies.

Freedom, anarchy and collaboration

> As he who lights his taper at mine, receives light without darkening me.
>
> (Thomas Jefferson)

Information age literature expresses a common belief that the ability to harness the power of collaboration is a product of the system's openness. Information has been important to human societies in all times and places. Further, networks themselves have always existed within human cultures. What is distinctive about this period of time is how 'new technology enhances the flexibility inherent in networks while solving the coordination and steering problems that impeded networks throughout history in their competition with hierarchical organizations'.[28] Networks distribute performance and share decision making; they are inherently flexible in their ability to add and subtract nodes without changing the fundamental organisation of their structure, permitting networks to maintain and enhance their value over time; finally, nodes enhance their relative importance by their ability to absorb and process information more efficiently. Technology linking nodes together in a seamless architecture, operating in an environment undisturbed by the clock routines of industrial organizations, seems to offer a radical emancipation of human thought and creativity. 'The growing accessibility of information technology puts the tools required to collaborate, create value, and cooperate at everyone's fingertips.'[29]

Indeed, the Internet has strong anarchical tendencies in many of its aspects. It is a 'place' without a 'space' and as such inherently resists control by territorial entities such as states. However, more fundamental seems to be the anarchical ideology that informs many such commentaries on the nature of the Internet. Classical anarchist philosopher Peter Kropotkin is cited by many as the basis for the theory of collaboration that powers the innovative aspect of the Internet. Both Rheingold and Eric Raymond of the Open Source Movement use Kropotkin's ideas on the ability of humans to cooperate without coercion in collective projects.[30] Tapscott and Williams argue seemingly anarchic principles as the basis for the new business economy that is being created by IT. Openness in terms of corporate boundaries and the movement of labour characterise the emerging business market of the twenty-first century; peering in the form of meritocracy typifies business transactions from Google's page-ranking features to eBay's trust measurement system between sellers and buyers. Finally, digital media's essential malleability enhances sharing between users allowing them to alter, remix, and repurpose content found on the Internet, thus creating new value from found objects. All of this enables the creation of truly global enterprises, building a 'planetary ecosystem for designing, sourcing, assembling and distributing products…'.[31]

Freedom, it is argued, gives networks their inherent power over hierarchical counterparts in this new technological structure. Companies developing open source software, such as Linux, Apache, or Firefox have inherent advantages in their ability to rapidly collaborate, develop, and fix software, outmanoeuvring their counterparts stuck in traditional industrial organisations.[32] Stallman famously points out that 'when I talk of free software, I am referring to freedom, not price. So think free speech, not free beer'.[33] However, Chris Anderson of *Wired* argues that 'free beer' is likely the destination of most products offered on the Web, 'everything the Web touches,' he notes, 'starts down the path to gratis.' Because the marginal cost of digital information on the Web is close to or actually at zero, 'free becomes not just an option but the inevitable destination'.[34]

Many besides Anderson have taken up Stewart Brant's argument that 'information wants to be free'. Some argue that information is like a life form itself, seeking the opportunity to determine itself. Information self-reproduces, virus like, spreading and persisting between individuals; digital information seems especially hard wired to mutate: 'digitized information has no final cut'; and depending on the context, is capable of perishing in its ability to degrade over time.[35] Clearly, the abundance of information that modern IT provides creates an environment where ideas compete against each other in terms of their perceived value to consumers. But this is also an environment in which consumers become 'prosumers' in their ability to shape and alter the information they receive and share.

Networks create this condition of abundance in their ability to circulate information rather than simply accumulate it.[36] It is this free circulation of information that confers on new forms of collaboration their distinct power. 'Given enough eyeballs, all bugs are shallow', the so-called 'Linus' Law' asserts that there is always somebody capable of solving every problem; the key is linking them up within a community.[37] While this example is typically used to argue for the superiority of open source software, it can be seen in other contexts as well. The revelation that the memo purporting to demonstrate that George W. Bush dodged the draft during the Vietnam war were quickly revealed to be forgeries through the collective analysis of the originals by bloggers watching the developing story. Careful observers noted clues in the font used on the memo which were unavailable at the time of its purported printing, and others familiar with such documents found discrepancies in its style.[38]

The complex nature of information generates the power that results from the circulation of ideas. While information may be akin to a life form, it is also an 'activity' as opposed to an actual 'thing' that can be possessed. Information is 'something that happens in the field of interaction between minds.... Information is an activity which occupies time rather that a state of being that occupies physical space, as is the case with hard goods.' As such, information is experienced rather than possessed, propagated rather than distributed.[39] Lash makes a similar argument noting that

> Capital accumulates. It already has some sort of order inherent to it. Information, on the other hand, circulates, it swirls, it bombards. Capital as assets accumulated means its production is found in specific zones. It may be exported for production in the third world. It may be internationalized... Capital, however, is not everywhere. You are not bombarded by it from billboards and in your own home. Information is in its nature much more anarchic than capital. Capital is regulated by the hidden hand of markets.... Information escapes the very logic of markets. It is everywhere at the same time for free. Information may be ungovernable.[40]

This notion challenges the belief that information can be owned in the same manner as capital. Copyright is used to protect the ownership of information and is the basis for ongoing discussions over rights to protect 'intellectual property'. However, some argue copyright exists for the protection of consumers by preventing unscrupulous publishers from appropriating the product of others. This worked well with books – in their innate physicality, they were difficult to produce and alter. As such, copyright protected the ability of publishers to make books rather than as distributors of ideas, 'the bottle was protected, not the wine'.[41] Copyright, it is argued, makes less sense when the circulation of ideas is more akin to pure thought than a physical product. Free software advocate Richard Stallman developed the concept of 'Copyleft' to address this shift in the distribution of ideas, such that consumers continued to benefit from their free and open circulation. Copyleft permits companies to charge money for their product, but extends the right to others to distribute and change it. Further, the rights continue to travel with the modified copies.[42]

Of course, such notions are not unchallenged, although free software and open source advocates argue that ultimately an 'evolutionary arms race' between closed source industries and open source networks ensure the latter will win.[43] Irrespective of this struggle, clearly such notions concerning the free circulation of all information will be difficult to accept in military and intelligence circles. The Robb-Silberman Commission investigating the intelligence process that underlay the supposed existence of weapons of mass destruction in Iraq found that 'the term information sharing suggests that the federal government entity that collects the information, "owns" it and can decide whether to "share" it with others. This concept is deeply embedded in the intelligence community's culture. We reject it.'[44] The CIA developed 'Intelink' in 1994 as a web-based information portal for intelligence information, and more recently 'Intellipedia' as a Wikipedia clone for the same purpose.[45] Furthermore, the CIA has also created a social networking site, 'ASpace' that mimics applications like MySpace and Facebook. Reportedly, Intelink was a relative failure as intelligence managers chose to withhold their most sensitive information, including operational details from the system. Intellipedia, reports suggest, is a relative success after the conduct of a 'marketing' campaign, although doubts persist within the intelligence community on its utility and its security.[46]

Intelligence agencies and military organisations resist the anarchist ideology of the Web for the simple reason that aside from its life-like and verb-like properties, information is ultimately also a relationship that exists within the mind. 'We assign value to information based on its meaningfulness', a relationship that can only be determined by an individual mind. In this relationship, scarcity and authority play critical roles. Because this is a human centric activity, the authority of the mind assigning the signification process altering data into information is important – some points of view are valued more than others. Second, some types of information are not abundant and thus have higher 'marginal value'.[47] As such, secrets retain their currency. Even Anderson notes that 'Information wants to be free, information wants to be expensive.... That tension will not go away.'[48]

In the differing values assigned to discrete bits of information, walls are created restricting its free flow. Most famously, the walls between members of the intelligence community have been blamed for the attacks of 11 September 2001.[49] It is an assumption that harkens to Linus' law – with enough analysts aware of all the facts, one would have spotted the 'bug'. It reasonably assumes that without barriers to the free circulation of information, any given analyst would have been able to ferret out all the relevant bits of information and piece enough to suggest the existence of a plot. Even assuming that such barriers can be lowered, a second issue concerns the ability of that analyst to compel others to listen to and believe the story thus crafted, and this is a question directly related to power.

Some have sought to address this problem. Rheingold argues that there is a difference between collectivism and collective action. The former involves coercion and centralised control whereas the latter is based on 'freely chosen self selection and distributed coordination' to achieve shared outcomes.[50] Nevertheless, as the authority of a particular observer's point of view is a crucial determinant of the meaning assigned to any specific relationship between datum points in a narrative, the relative power of that observer as opposed to all others weighing the significance of his narrative is crucial. Jaron Lanier has criticised notions like Rheingold's as 'Digital Maoism'.

> The beauty of the Internet is that it connects people. The value is in the other people. If we start to believe that the Internet itself is an entity that has something to say, we're devaluing those people and making ourselves into idiots.[51]

In other words, real people as opposed to a mythical collective exert the real power. Leaders in open source projects

> are often described as democratic individuals. But in practice, key developers tend to see their positions as a licence to make unilateral decisions. These leaders fought hard and often paid personal costs to achieve their status. Sharing that power with others runs directly against their own aims and ambitions.[52]

Others have also noted this philosophical problem at the heart of digital anarchism; Castells, Tapscott and Williams, and Lash have all commented on the underlying power that specific individuals or even 'nodes' within networks can have in shaping how information circulates on them.[53] Still, Tapscott and Williams counter that 'the basic rules of operation (for open source projects) are about as different from a corporation command and control hierarchy as the latter was from the feudal craft ship of the pre-industrial economy.'[54]

Irrespective of whether this is true or not, it is clear that the state continues to operate in a 'command and control' hierarchy with regards to information, especially the most sensitive sort. The 'walls' protecting sensitive information from hostile eyes as well as friendly ones on social networking sites arise as much because of the explicit value of the information as from the uncertainty about its final value. Indeed, as the significance of information depends on the meaning assigned to the datum and the arrangement of facts into a coherent narrative, the ultimate significance of any piece of information is essentially unknowable in advance. As such, agencies tend to overclassify material in order to manage the risk of inappropriately revealing what may need to be kept secret. This issue is magnified by the inherently fluid and mutable nature of digital content.

Intelligence and military operators obviously desire to protect sources and techniques, just as law enforcement agencies seek to avoid compromising the jurisprudence of criminal cases under investigation. However, all need to cooperate in order to deal with the complex pan-jurisdictional nature of the global War on Terror. A 2008 US government report discussing the possibility of developing an 'information sharing environment' uniting disparate governmental communities develops one solution in the form of an 'authorized use standard' which would protect the conflicting demands to keep some things secret and share others.

> A regime operating under an 'authorized use standard' would enable an appropriately credentialed official to access any information in possession of a US government agency based *not* on an application of legal and policy requirements currently in effect ... but rather on whether the official has the proper mission based or threat based permission to access that information. The key determination would be whether that permission was for 'lawful purposes' and the process for making that determination would be established in consultation with the Privacy and Civil Liberties Oversight Board.[55]

Such permission would not be straightforward to determine, however. Considerations would have to be based on the legal authority of each agency involved, their specific missions, the sensitivity of the information and how it would be used, and that any sharing would be consistent with constitutional principles, statues, presidential executive orders, regulations, the user's authorised mission, *and* the mission of his/her agency. Such a determination would apply only to US persons and permanent residents and continue to require additional authorisation where required by law.[56] It is a process that is complex, time consuming, and one which reinforces the relational aspects of information that create barriers to its dissemination in the first place. In no way does it create an environment which replicates or even mimics the anarchical distribution of information evident on the Internet.

One final criticism arrayed against the digital anarchism of the open source/free software movements is the ability of networks to exclude as well as to include.

> The true value of the informal group of co-developers and users often is only revealed after a developer has abandoned his project. And the loss can be painful as the social and professional bonds that the developer acquired may not survive his or her 'defection'.[57]

The risk of such exclusionary tactics is considerably greater in the sphere of military and intelligence information sharing. The example of New Zealand's stormy relationship with the United States is a case in point. Following the decision of the David Lange government in 1985 to ban port visits of nuclear powered/armed US Navy ships to New Zealand ports, the US suspended its obligations to the country under the ANZUS pact. This had a dramatic effect in terms of information sharing between the two countries, although Hager alleges that intelligence cooperation between New Zealand's Government Communications Security Bureau and the American National Security Agency was largely unaffected given the important value of the information it provided.[58] Following decisions to abstain from cooperating with the US in the invasion of Iraq and, later, missile defence, the US briefly exercised similar policies against Canada.

The aporetics of information sharing

As can be seen, information exchange between nations in the digital era has become progressively more complex. A whole series of instructions governing information exchange, especially involving information systems and networks, has been released by a variety of government bodies since the early 1990s. In a 2004 memo, Steven Cambone wrote that

> Our ability to share in a timely manner will determine our ability to leverage our unmatched capabilities. In order to accommodate new and rapidly changing demands to share information and to handle it in a secure electronic environment, information that has been determined releaseable through established foreign disclosure procedures to foreign networks ... shall be marked 'Releaseable to USA with the applicable trigraph...'.[59]

The depth of regulation controlling the storage, use, and dissemination of classified information illustrates best the clear differences in how information is formally treated within governmental bodies. In 2005, US Army Chief of Staff, General Peter Schoomaker drew explicit attention to the crossover between these two domains in concerns regarding the use of blogs by soldiers: 'The enemy reads our open source and continues to exploit such information for use against our forces.'[60] Such concerns have been raised repeatedly against the use of blogs, social networking sites like MySpace, and the posting of combat videos on video sharing sites like YouTube and LiveLeak by service members. While official policy is ostensibly to permit as much latitude as possible,[61] the impact on soldiers using blogs has been chilling.[62]

Of particular interest is the treatment of unclassified information by some of this policy. For example, US Army regulations on operational security note:

> (3) Critical information is information that is vital to a mission that if an adversary obtains it, correctly analyzes it, and acts upon it, the compromise of this information could prevent or seriously degrade mission success.
> (4) Critical information can *either* be classified or unclassified. Critical information that is classified requires OPSEC measures for additional protection because it can be revealed by unclassified indicators. Critical information that is unclassified *especially* requires OPSEC measures because it is not protected by the requirements provided to classified information. Critical information can also be an action that provides an indicator of value to an adversary and places a friendly activity or operation at risk.[63]

As the instruction points out, unclassified information poses potentially as much a risk as classified material. Indeed, a whole series of new classifications for information have emerged to complement the classic Secret and Top Secret classifications including 'For Official Use Only' (FOUO), 'Sensitive But Unclassified' (SBU), and 'Controlled Unclassified Information' (CUI) which all control information that had previously been publicly available.[64] This trend complements Nico Stehr's observation on the surprising resilience of borders in the information age: 'The world may be opening up and the circulation of fashions and goods and people becoming more intense, but differing convictions as to what is "sacred" still create insurmountable barriers to communications.'[65] In this regard, national security information is as sacred as any religious tenet.

The baroque nature of information classification, however, further complicates its management. In his 2004 Congressional testimony, the DoD Director of Information Security Oversight admitted that 'information that should not be classified is increasing, in violation of Executive Order 12958'.[66] A GAO report from 2006 on DoD's management of classified information found that classification management training in the US military was insufficient at many levels, that they were not uniformly following established procedures for classifying data, that there was inconsistent treatment of similar types of information within the same documents, and material marked as classified often did not meet the established criteria for doing so.[67]

Clearly, the portability and mutability of digital information compromises the willingness to share digital data. Tools designed to share information can be turned against their own users. As Scott Lash has pointed out, IT itself is the source of information overload given the ease of generating and disseminating digital information, along with the proliferation of sources for information.[68] This inherently creates the anarchical spread of information. Sophisticated search technology is now capable of imposing some order on this ocean of detail, allowing the persistent researcher to winkle out individual and innocuous bits of information that might form patterns, revealing larger issues, intentions, or compromise projects.

As information classification standards proliferate globally, the management of relations between intelligence and military organisations itself becomes more complex. Each agency manages a web of relationships with counterparts in other countries with differing degrees of openness.[69] As each country has different needs and regulations determining what is collected and how it is stored, the standards on which information is classified between nations can vary widely. In a digital age, this can add considerable complexity to the process of sharing information. 'Data models' specify the nature, organisation, and relationships between fields in information databases. These models can differ significantly in the diverse information products used by various military and intelligence services.

Some argue that such digital anarchism may be an 'early transient phenomenon', that a 'more stable configuration' will ultimately emerge where worker and managerial autonomy is constrained within the digital environment.[70] Several recent studies have argued for the possibility of a more dystopian future with digital technology.[71] However, these transformations within our society may be larger than the technology that presents these opportunities to reassert control.

> What is characteristic of social movements and cultural projects built around identity in the Information Age is that they do not originate within the institutions of civil society. They introduce from the outset, an alternative social

> logic, distinct from the principles of performance around which the dominant institutions of society are built.... The strength of identity based social movements is their autonomy vis a vis the institutions of the state, the logic of capitalism, and the seduction of technology.[72]

In other words, the possibility of control may escape those forces best positioned to implement it. While this may be cause for celebration in some corridors, Valovic notes that 'digital culture cannot prize the anarchic and chaotic qualities of the Internet above all else and yet expect some kind of pluralistic cultural system or even model of governance to arise from those qualities'.[73] The digital environment may ultimately come to resemble the international environment in its worst anarchical qualities.

Here we finally end up with the aporetical relationship between the organs of the state, the military in particular, and the emerging informational society. In his study of the relationship between these two forces, Everard concluded that the state was here to stay because of its connection to both the formation of social identity and its monopoly of violence.[74] Both these roles are placed at risk by the rise of networks. But it is the relationship between the military and the informational society as it affects the sharing of information that concerns this study. The network structure of new social groups has clearly led to a quantitative and qualitative increase in collaboration and innovation. The steady spread of new applications on the Web continues to capture public imagination; military interest in taking advantage of this phenomenon is evident.

The importance of secrecy in protecting information places clear constraints on the ability of militaries to use technology to the same advantage as those groups which are rising on the Internet. This division is all the more intriguing given the origin of the Internet as a military project. Paul Baran, the Rand analyst who authored the first conceptual discussions of what would become the Internet, noted that the technology would assist as a means of keeping military secrets safe from foreign spies in peacetime as well as protecting communications during war.[75] Castells agrees with Everard that the state will survive the transformation in human society he foresees approaching, however, he disagrees that sovereignty will survive intact because of the impact on society of emerging forms of networked organisation. Although he does not consider the implications of this evolution for international relations in any great depth, he does advance two conclusions. First, that we might expect greater amounts of multilateralism in geopolitics. At the same time, he also expects that geopolitics will be 'increasingly dominated by a fundamental contradiction between the multilateralism of decision making and the unilateralism of the military implementation of those decisions'.[76] Here again, we see the aporia between the inherent impulse to work together that is emerging from shifts in culture, the economy, and technology with the need to wall off information, the release of which might damage national security. Just as the technology that enhances the ability of humans to cooperate together is emerging, we may expect to see the opposite trend within the military sphere – the growing difficulty of achieving the same. That is the subject of the next chapter.

1 US military primacy and the new operating system

Two issues currently play dominant roles in shaping the current international landscape. Processes commonly referred to by the label 'globalisation' are affecting every area of the world through environmental modification, electronic communications, financial shifts, and the evolution of a worldwide civil society. Juxtaposed against this multidimensional globalisation is US military primacy. In 2004, the United States spent $466 billion on defence; the next largest spenders were China and Russia, at $65 billion and $50 billion respectively.[1] These twin developments, one generalised across the planet and the other specific to the US, will interact in complex ways as the world responds to US military primacy and as an increasingly globalised environment compels political, economic, humanitarian, and military engagement between states.

The conjunction of these two issues underlines the political applicability of US military power. This became apparent early on in the War on Terror, when then US Secretary of Defense Donald Rumsfeld indicated that military imperatives would take precedence over diplomatic considerations in constructing a 'coalition of the willing'. According to Rumsfeld, the US must 'avoid trying so hard to persuade others to join a coalition that we compromise on our goals or jeopardise the command structure. Generally, the mission will determine the coalition; the coalition should not determine the mission.'[2] Such political bravura has now receded significantly as the US actively courts the shrinking number of partners willing to work with it in dangerous missions. However, as the only leader capable of mounting large, complex, and global operations, how the US conducts its future military operations will shape how others conduct theirs.[3] In IT terms, American doctrines will dominate military operations in the same way that Microsoft's Windows operating system dominates computer programming.

American hegemony and military primacy

Military power is only one aspect of America's hegemonic position, and it is by no means always relevant. The greatest source of US strength is found in the ideological sway it holds over much of the world; its inherent 'soft power', as Joseph Nye characterises it. The United States exercises an ideational authority unlike that of any other state. Despite the United States' ideational influence, the extent of its control over the global economy is open to question. Moreover, a rising tide of anti-Americanism is challenging American soft power. The status of American military power, however, is largely beyond question, both rhetorically and in practice.[4]

The basis of American military primacy has been described in terms of 'command of the commons'. The 'commons' are those areas over which there is no national jurisdiction (most obviously, the sea and outer space) and those areas where military control is difficult to enforce. These areas can be used by any actor possessing the requisite capability. But because of the ubiquity of America's military power, as opposed to the 'niche' and localised roles played by other states, the United States is able to exploit these areas more effectively in pursuing its military ends. More importantly, it may deny the commons to others. Wresting command of the commons from the US would require a generalised war, which is clearly currently beyond the capabilities of any other state. Command of the commons in its essence gives the US global agency, a privileged global ability to act in other words. When the US confronts its enemies in their own specific areas of local control, they will already have been greatly weakened through diplomatic, economic, and moral isolation, and through stand off military strikes from air, space, and the sea.[5]

Command of the commons is enhanced not simply through the comprehensive nature of American military might, but also through its capacity for a global approach to operations. No other state possesses a comparable worldwide network of military outposts in friendly states, which provide logistics support for operations distant from the US homeland. The wide-ranging exercises that US forces conduct with the armed forces of allies and security partners also enhance American familiarity with the operational characteristics of international military actors, and with diverse operating environments. Finally, no other state organises its military activities on a global basis, as the United States does in the form of its Unified Command Plan (UCP). The UCP enables the American military to 'develop responsive war plans that can generate significant combat power in far corners of the world on relatively short notice'.[6]

The globalised nature of American military power does not obviate the need for allies and other security partners, as every US National Security Strategy has pointed out.[7] Although some have challenged the notion of unipolarity on this basis, it is the case that the strong 'have more ways of coping' than weaker powers.[8] The point here is that, at present, America's overwhelming military power provides it with options to structure the world that other states do not possess. In previous eras, this type of dominant power would have been of such concern to other states that it would have given rise to alliances, arms races, and outright political and military confrontation. There is speculation that the European Union (EU) may evolve as a potential counterweight to American hegemony, or that China will in time become a potential peer competitor. However, the fact that war between the major states is now largely 'unthinkable'[9] suggests that American power does not threaten the core interests of potential rivals in the way that the rise of Spanish, French, German, or indeed Spartan power did in the past. Concern over America's power centres on more generalised unease about global issues confronting all states, such as climate change, religious extremism, and cultural domination in all its varied forms. The nature of these challenges has limited reactions to the scope of US power to considerations of restraint and 'socialisation' in order to keep American behaviour within acceptable boundaries, much as one would deal with a friendly but unruly dog.[10] The issue for other nations is not wars of re-balancing, but how to engage American power: prodding it into action here, restraining it there. Thus, while the US seeks to shape the future of the world, other states seek reflexively to shape America's engagement with it. In the present global society of states, everyone has their own 'special relationship' with the US.

Allies and dominance

The public falling out in 2003 between the United States and some of its allies, particularly France and Germany, caused some to wonder whether American hegemony might be declining. Unlike in 1956, when the US was able to force France and Britain to back down over Suez, in the post-Cold War environment of 2003, Washington was unable to make its allies modify their policies. Indeed, as the dispute went on, each side became more intransigent. 'What this shows', argued Christopher Layne, 'is that it is easier to be number one when there is a number two that threatens numbers three, four, and five, and so on. It also suggests that a hegemon so clearly defied is a hegemon on a downward arc.'[11]

Yet as the French historian Raymond Aron has noted of another hegemon's decline, 'a change from *Pax Britannica* to the *Pax Americana* did not involve a change of universe, and pride, rather than the soul itself, suffered'.[12] Aron's observation points to the surprising absence of competition between the United States and Britain as they exchanged roles in the twentieth century. But it also bears some relevance to the absence of military competition between America and its Cold War partners. One might point to the process of military 'de-globalisation' that took place throughout the 1990s.[13] While American military spending fell somewhat in the early part of that decade, it has since recovered to the levels of the 1980s. At the same time, no state has responded in kind to American spending, and none has sought to challenge US dominance in key areas of military technology such as electronic warfare, intelligence, and surveillance. No peer competitor, whether China or Europe, has emerged in the military realm since the end of the Cold War, and no state seems likely to challenge the US militarily in the near future. Given the huge disparities in power between the US and China, massive increases in China's military budget would be necessary to develop the kind of power projection capabilities the US currently enjoys. Furthermore, such enormous changes would take years to mature to the level of operational proficiency that the US currently exercises.

A neutered Europe 'unable to focus its latent military power',[14] comprised of states incapable of fighting among themselves, Layne argues, has long been the goal of American policy.[15] If this is so, then it is at odds with America's declaratory policy throughout the Cold War, and its continued irritation with the lack of European burden-sharing since 1991. Still, the creation of a strategic environment dominated by American power has been part of US security policy since the end of the Cold War. In 1992, a draft copy of the still-classified *Defense Planning Guidance* was leaked to the *New York Times* and the *Washington Post*. As the *Post*'s Barton Gellman reported:

> The central strategy of the Pentagon framework is 'to establish and protect a new order' that accounts sufficiently for the interests of the advanced industrial nations to discourage them from challenging our leadership while at the same time maintaining a military dominance capable of deterring potential competitors from even aspiring to a larger regional or global role ... 'we will retain the pre-eminent responsibility for addressing selectively the wrongs which threaten not only our interests but those of our allies or friends, or which could seriously unsettle international relations'.[16]

This was a first attempt at reformulating American security policy to take account of the changes accompanying the end of the Cold War. Some argue that it was based on an honest attempt to reassess the doctrines that would guide American action abroad, and what America could expect in terms of cooperation from partners no longer existentially threatened as they had been throughout the Cold War.[17] Gellman noted that the 1992 document was not a revolutionary departure from traditional American policy, which had sought to ensure that no one power dominated any key region, placing it in a position to alter the global balance of power.[18] And indeed, the leaked document did refer specifically to the necessary role of allies and coalition partners, noting their 'considerable promise' in assisting America to further its interests abroad.[19] Additionally, some have argued that the 1992 document was in keeping with 'American exceptionalism', the notion that the United States always uses power benevolently. Some have noted this aspect of America's 'myth of invincibility', arguing: 'According to this faith, American global power is limited by its own political scruples and humanitarian self-restraint.'[20] Despite this, the document barely conceals its scepticism that such cooperation would be easy to orchestrate, or would be there simply for the asking: instead of relying on its own system of alliances, the US 'should expect future coalitions to be ad hoc assemblies', and 'should be postured to act independently when collective action cannot be orchestrated'.[21]

The policy did not withstand the withering criticism directed at it from both the media and America's allies; the language of *Defense Planning Guidance* was altered to make it more acceptable, and it seemed to be relegated to the status of a footnote in US security policy. However, the emergence of George W. Bush's first *National Security Strategy* in the post-9/11 environment strongly recalls the words of the discarded 1992 *Guidance*.[22] Shortly before its publication, Bush noted in his 2002 address to the graduating class at the US Military Academy at West Point that 'America has and intends to keep military strength beyond challenge – thereby making the destabilising arms races of other eras pointless'.[23] The undertones of 1992 in subsequent US strategic policy, and the participation of several personalities from the first Bush administration, including the original document's author, Paul Wolfowitz (who became Deputy Secretary of Defense in 2001), linked the two policies.

It seems that the quest for military supremacy remained part of Pentagon policy post-1992,[24] as shown by the development of the concept of 'Full Spectrum Dominance' during the mid-1990s. First articulated in the 1995 document *Joint Vision 2010* (JV2010), Full Spectrum Dominance was supposed to enable the US 'to dominate the full range of military operations from humanitarian assistance, through peace operations, up to and into the highest intensity conflict'.[25] Here was the articulation of a policy that called for American pre-eminence across the full span of military operations, not just in traditional conventional force-on-force engagements. The goals of 1992's *Defense Planning Guidance* might officially have been renounced, but they persisted as the sub-text to the development of the US military's response to the Revolution in Military Affairs (RMA). This strategic approach to novel military technology and new forms of organisation is clearly apparent in the erstwhile Office of Force Transformation's definition of military transformation as:

> A process that shapes the changing nature of military competition and cooperation through new combinations and concepts, capabilities, people, and organisations that exploit our nation's advantages, protect against our asymmetric vulnerabilities to sustain our strategic position which helps underpin peace and stability in the world.[26]

It is worth recalling Wolfowitz's observations, in a 2000 edition of *The National Interest*, on America's remarkable success in forming coalitions. According to Wolfowitz, this had been achieved not by 'lecturing and posturing and demanding', but by:

> demonstrating that your friends will be protected and taken care of, that your enemies will be punished, and those who refuse to support you will live to regret having done so. It includes lessons about the difference between coalitions that are united by a common purpose, and collections of countries that are searching for the least common denominator and for easy ways out of a problem.[27]

In the same issue, Robert Kagan and William Kristol, commentators closely associated with the so-called neo-conservative movement, raised similar themes in their article, entitled 'The Present Danger':

> Those alliances are a bulwark of American power and more important still, they constitute the heart of liberal democratic civilisation the US seeks to preserve and extend. Critics of a strategy of American pre-eminence sometimes

> claim that it is a call for unilateralism. It is not. The notion that the US could somehow 'go it alone' and maintain its pre-eminence without its allies is strategically misguided. It is also morally bankrupt.[28]

Of course, this was not the first time that spokesmen for a pre-eminent power expressed such sentiments. Nearly 2,500 years earlier, Pericles extolled the exceptionalism of Athens and its generosity towards its allies:

> it is only the Athenians who, fearless of consequence, confer their benefits not from calculations of expediency but in the confidence of their liberality. In short, I say that as a city we are the school of Hellas; while I doubt if the world can produce a man, who where he has only himself to depend upon, is equal to so many emergencies, and graced by so happy a versatility as the Athenian. And this is no mere boast thrown out for the occasion, but plain matter of fact, is proved by the power of the state acquired by these habits.[29]

While we may debate the limits to and constraints on American power, pointing to loosened control over global shifts of capital, growing anti-Americanism, the potential rise of new 'balancing' powers like China or the EU, no actor shares the will and capacity to act globally that is at the heart of American military primacy. Given the absence of investment by other states and institutions in building their military capability, US military pre-eminence is likely to remain unchallenged, at least in the near term. This singular capacity to command the commons, to act militarily at a global level as opposed to every other power's limited niche or local capabilities, challenges the very nature and need for alliances despite the apologetic language inserted, de rigeur, in national security strategies. It is this capacity, possessed by a singular nation, that prompted Singaporean diplomat Kishore Mahbubani to call in 2005 for a 'new contract between America and the world':

> There needs to be an open and candid discussion, involving all sections of humanity, on the nature of the world order that will be realistically supported by America, the major powers, the weaker states, and the intelligent human community.[30]

It is a plea that can only be termed reasonable in the context of the vast disparity of power enjoyed by a single state compared with the rest of the world. However, the nature of globalisation and the risks it entails for all states ensure that enough common ground on which to base such a contract is unlikely to emerge quickly.

Globalisation, security, and risk

Some have portrayed the split between the United States and the Franco-German axis in 2003 as a strategic sea change.[31] Of course, NATO has often been on the brink of crisis, whether over basic strategy, nuclear weapons, *Ostpolitik*, or burden-sharing.[32] The parting of company between erstwhile friends in this instance occurred over issues located far from Europe, and points to the changing nature of the transatlantic partnership, confronted by the challenges of failed states, nuclear proliferation, and global terrorism. If unity on direct threats to national existence was hard to achieve, what hope can there be for unity over less immediate and more geographically distant issues?

The very nature of globalisation points to a complex future wherein insecurity is inextricably bound up with the promise of progress. The complex web of interdependent and cross-cutting relationships that make up globalisation not only makes its precise definition difficult, but also leads to considerable uncertainty in terms of its overall long-term social, political, and economic effects.[33] Globalisation is inherently political in its tendency to produce both winners and losers depending on the nature of this complex interplay of variables. As each globalised relationship will produce variable outcomes for every participant, it is impossible to blandly characterise the overall process as either 'good' or 'bad', 'stabilising' or 'divisive'. As such, globalisation is by its nature ambiguous, and thus a source of insecurity even as it generates opportunities; it is at once enabling and disempowering.[34]

Globalisation permits unstable regions to have strategic impact far beyond their local areas. Such 'zones of war' produce 'leaking misery' in the form of terrorism, crime, and refugees (both political and economic) heading for 'zones of peace'.[35] The result is intervention in failed states involving operations between paramilitaries, conventional forces, and NGOs, undertaking a variety of operations including nation-building, humanitarian assistance, counter-insurgency, indigenous force training, and outright combat – what the US Marine Corps describes as a 'Three Block War'.[36] In sum, globalisation produces an inherently complex security landscape defying any single solution around which international agreement can easily crystallise. This landscape will politically mobilise a multiplicity of interests stretching across these zones of peace and war, further complicating efforts to find common ground.

An explicit example of the cross-cutting nature of globalisation is found in the role of global communications. The ability of ordinary individuals to inform themselves on international issues has contributed to the emergence of a 'global citizenry', capable of monitoring state action and insisting on the application of universalised ethical norms to any state's policy.[37] Videos, often filmed with cellphone cameras, documenting torture of prisoners or other injustices have frequently found their way onto the Internet through sites like YouTube and from there onto traditional media such as network news, or less traditional ones in the form of blogs. This has produced what Moisés Naim calls 'the YouTube effect' where even momentary clips can gain an enduring presence thanks to the ability of IT to propagate information on a global basis.[38]

However, the same technology also permits those less committed to universalised notions of human identity to exploit differences in forms of justice, and to provoke violence between communities. The globalised riots and demonstrations against the negative portrayal of the Prophet Mohammed in cartoons in an obscure Danish newspaper in early 2006 point to the fragmenting effects that global communications may have. All of this points to the anarchical nature of information raised in the Introduction, which will further complicate the management of international issues. Just as a YouTube video can be used to inform the global citizenry, it can also be used to misinform and mislead for clear political effect. Naim believes that the 'wisdom of crowds will ultimately correct "photoshopped" pictures, staged videos, and other digital media that have been mashed together'.[39] Indeed, during the 2006 war between Israel and Hezbollah, bloggers were successful in outing a number of manipulated photos, some of which received wide publication.[40] However, the anarchical nature of information will raise significant issues on how reliable or trustworthy these reassessments are. Second, reanalysis will always come much later, the detection and scrutiny of data taking longer than the initial political effect. In this environment, YouTube videos can be like fire-and-forget missiles.

As Lawrence Freedman has noted, 'a world in which threats are real enough but do not come from other Great Powers is bound to ask different questions of an alliance than one which is focussed on deterring or fighting a major war'.[41] The questions that will be asked of any partnership of powers will revolve around the uneven sharing of risks between these powers. What is striking about this condition is the necessarily subjective context in which consideration of potential policy alternatives takes place. The uneven nature of risk implies highly contextualised and individualistic definitions of what constitutes the 'correct' course of action. Modern democratic societies, politically mobilised by considerations of peace and war, are particularly prone to such debates given the 'risk' that military operations represent.

The question of risk as a fundamental aspect of modern society has been discussed extensively within sociological literature.[42] 'Risk society' emerges from the critique of the idea of progress. The notion of reflexive modernisation is the process by which society recognises that there is a price to be paid for all progress – that all actions have unintended consequences, whose nature often cannot be anticipated in advance. Because all actions carry the price of uncertain outcomes, risk assessments come to dominate all decisions regarding what action should be undertaken.[43] As Anthony Giddens reminds us, the notion of risk has always been present in human society in relation to natural forces that unfold in unforeseen ways. As our ability to shape our own environment developed, however, modern society began to encounter 'manufactured risks': man-made hazards as threatening as any in the natural world.[44] The nature of 'modern' society is to try and foresee and thus control the future consequences of human action. However, the consequences of nuclear disasters, climate change, the global spread of disease and invasive flora and fauna facilitated by modern transport, financial collapse facilitated by electronic currency speculation, and the effects of emerging technologies such as genetic engineering and nanotechnology, are so great and widespread as to be largely beyond the control of any single individual, group, organisation, or state.

Risk defines itself in terms of its unpredictability and the uncertainty of cause and effect, thus removing it from the rational realm of scientific determination: one can speak only of probabilities.[45] Further, these risks

> can no longer be limited to certain localities or groups but rather exhibit a tendency to globalisation which spans production and reproduction as much as national borders and in this sense brings into being supra-national and non-class specific global hazards with a new type of social and political dynamism.[46]

This dynamism is raised in the context of assessments on the probabilities of hazard as defined by the opinions of 'experts'; such 'social dependency on institutions that are alien/obscure/inaccessible to those affected raises issues of trust and credibility'.[47] As risk is uncertain, it is inevitably politicised because of the varying impact it has on various social interests, each deploying its own experts and spokespersons. Therefore, the 'existence and distribution of risks and hazards are mediated on principle through argument'.[48] In its nature, risk is, therefore, socially constructed and articulated by the values and interests of those perceiving the risk.[49] The uncertainty that surrounds risk politicises it in terms of 'cover-ups' and 'scare-mongering'.[50] Debates over terrorism and WMD, both pre- and post-9/11, have exhibited both these characteristics.

There is a subtle link between informationalism and risk society. Nico Stehr points out that 'if knowledge is the main constitutive characteristic of modern society, the production, reproduction, distribution and realization of knowledge cannot avoid becoming politicized'. The same processes which are delegitimising the large monolithic social institutions are responsible for generating this effect. As such, risk society is also a product of informationalism.[51]

Harvard political scientist Michael Ignatieff inadvertently uses the language of the risk society when he speaks of the inevitable 'political and moral debris' that accompanies all military action.[52] Indeed, war is the 'ultimate' in risk management:

> Air strikes are vulnerable to the vagaries of the weather, incorrect intelligence and the malfunction of sophisticated computers and guidance systems. Air crews might make errors of judgement under applicable rules of engagement, especially if they are engaged by the adversary's air defence weapons. As demonstrated in Kosovo, there can be accidents and mistakes even when targeting has been subject to meticulous planning and careful consideration.[53]

War, of course, is a highly politicised phenomenon, particularly within democratic societies, which must be convinced of the appropriateness of military action before sanctioning it. In these societies, risk assessment is nowhere more evident than when multinational military action is contemplated. In such circumstances, debate over war is not simply polarised, but also 'globalised'. The question of just how much of a threat Saddam Hussein represented to international peace and order in 2002–2003 was largely framed within the context of a global debate on risk. The legions of prewar assessments conducted by state intelligence agencies, international bodies, and NGOs, together with phalanxes of informed experts on all sides of the question, was as much an orchestrated campaign as the military one that followed. That so many well-informed analyses later turned out to be so incorrect illustrates the subjectivity of risk assessments in relation to war.

Power and discourse among nations

Globalised debates on the correct course of action are nothing new to the field of international relations. IR is essentially engaged in an 'unending search for an understanding of the relationship between order and justice',[54] but is particularly challenged by conflict between different social and political ideals. Thus, the balance between order and justice is a timeless and discursive process. The end of the Cold War made this deliberation all the more pressing, as well as more difficult. The emergence of 'human security' as a focal point for international action, the rise of environmental concerns, and the empowerment of new voices attending the process of globalisation, whether in terms of debate over 'fair trade' or *jihadist* critiques of Western modernity, all added to what was already a complex agenda for establishing international social justice. But the end of the Cold War also gave rise, initially at least, to the hope that some solution might be found in terms of a 'New World Order' amongst the 'Free World' and newly democratised nations in an embedded liberalist epistemological community. This collective order of democratic states sharing common human values was ultimately unable to achieve unanimity on many issues of 'governance'. This should have come as no surprise: if, as Freedman suggests, 'strategy is the art of creating power',[55] then any strategy contingent on collective action must necessarily prioritise compromise as a key enabling condition of that strategy. Compromise, however, often means suboptimal results, shown particularly by the lack of action over Darfur, and the nearly botched NATO operation against Serbia in 1999. The fading of the optimism that had initially greeted the end of the Cold War is thus more representative of the enduring 'clash of moral, national, and religious loyalties' reflecting 'the plurality of values by which all political arrangements and notions of the good life are to be judged'.[56]

Despite the pressure for action in Darfur, a resolution of that crisis seems as remote as ever – even in the face of numbers of killed, wounded, and displaced many times more significant than those affected by the attacks of 9/11. Calls for action have been met with more diplomatic posturing by even those most supportive of such policies. Certainly, this points to the fact that human security was always simply an illusion nurtured in the euphoria that accompanied the end of the Cold War. The plain reluctance of developed states to place their troops at risk in support of humanitarian missions indicates the hollowness of the values supposedly underlying their commitments to human security.[57]

The comparison between action on Iraq and non-action on Darfur reveals just how hollow this supposed commitment is. In each case, the United States has provided an important leadership role. With Darfur, the US has gone so far as to label what is happening there as genocide, in the hopes of both shaming other states into action, and to establish a legal and normative basis for intervention. However, where the US was seemingly willing to conduct Iraq operations unilaterally if necessary, an intervention force for Darfur remains stalled at the diplomatic level, and whatever funding is available is clearly far less than what America is willing to spend in Iraq.[58] Europe, too, bears considerable blame in this matter. Where it was defiantly unwilling to intervene in Iraq, on human security issues supposedly more in keeping with Europe's ideological *Weltanschaung*, it has soft pedalled its reluctance to intervene in Darfur. Nor have the crowds across Western Europe, so opposed to war in Iraq, materialised to demand action in Darfur.

Inaction in this case has more to do with hard-edged compromises to political principles than any moral failing. The discursive nature of strategic decision making, especially when those decisions are shared amongst many partners, requires a complex mixture of cooperation, confrontation, and competition in the process of hammering out the compromises of policy details.

> Interdependence of decision making means that effective strategy is based on the relationships involved and the opportunities it provides the various actors. It is necessary to anticipate the choices faced by others and the way your action shapes those choices.[59]

Naturally, compromise is an important currency in political relations, even in highly adversarial ones. The credibility of action can come to suffer if too much compromise is made, thus undermining the objectives sought. Strategic compromise may ultimately result in compromised operations.

As Ignatieff points out, in Kosovo Slobodan Milosevic took on military forces greatly superior to those he himself wielded, and nearly won.[60] In that conflict, Milosevic enjoyed the advantage of being a unitary actor confronted by a complex coalition of powers only loosely held together by a broadly defined common objective. NATO fought under considerable constraints, which Yugoslav forces did not share. Intense political pressure was applied to minimise casualties (friendly, enemy, and civilian), minimise attacks on civil infrastructure, and rapidly halt ethnic cleansing.[61] The tensions between NATO's wartime objectives were a product of the tangled negotiations that ultimately brought the alliance to the first use of force in its long history. Potential Russian and Chinese vetoes meant no Security Council mandate was in place, nor was NATO able to agree on a single legal basis for the war, with each member state applying various legal and political justifications. The UN itself was placed in a difficult situation: it wanted to see Kosovars protected from Serbian attack, yet at the same time needed to protect its authority in establishing the legitimate use of force according to the Charter. Kofi Annan, the UN Secretary-General, was ultimately forced to split hairs, noting that 'it is indeed tragic that diplomacy has failed but there are times when the use of force may be legitimate in the pursuit of peace'.[62] Political compromises extended into the cockpit, where complex choices were faced between the need to protect pilots by flying at high altitudes, above the range of all but the largest Yugoslav air defence weapons, and the need to protect civilians from inadvertent air strikes by flying much lower so that pilots could properly identify their targets.[63]

Finally, there were differences in how NATO partners interpreted the laws of armed conflict. States that had ratified the Geneva Conventions of 1977 interpreted Protocol One, prohibiting 'excessive' civilian casualties, as 'treaty law', setting a very high standard of practice that meant virtually any civilian casualty was excessive. The US had signed the Geneva Convention but had not ratified it, and thus interpreted Protocol One as 'international customary law' determined by the benchmark of how the United States has traditionally conducted air operations. This permitted a much looser standard with regard to what constituted excessive civilian casualties. The operational impact of this intra-alliance conflict of interpretations meant that some targets were off-limits, not only to specific nations but also to NATO in general. In practice, this generated two separate Air Tasking Orders, one for the alliance and a second for the US alone.

These differing interpretations of what constitutes 'excessive' recall the issue of risk discussed above. Excessive casualties are a 'risk' of any air operation, and place states and pilots not only in moral, but also legal, hazard. However, like risk, 'excessive' is a socially constructed definition, not strictly resolvable objectively. How it is defined will depend on a complex grouping of factors, including culture, history, national psyche, and military doctrine.[64] Unpredictable elements, such as public opinion, will also play a significant role. Publics under direct threat will obviously define 'excessive' in different ways than those who are disconnected from the impact of war.

In standardising and regulating international behaviour, the expectation is that law, especially in combination with high-precision weaponry, will reduce, if not eliminate, the moral and political hazard of engaging in risky interventions. However, as has been shown repeatedly in military operations throughout the 1990s, the fact that Western forces may hold their actions accountable to high legal standards does not mean that their opponents will do so.[65] The use of hostages to deter military strikes or lower public morale is increasingly common on the modern battlefield, as are strikes against civilians as proxies for military targets, thus increasing the political and moral complexity of military responses.

Power is a requisite for action, and if strategy is the art of creating power, then compromise and cooperation ultimately are important aspects of it. As Freedman continues, however, power is a relative concept, existing only as it is recognised by others, whether that recognition devolves from simple authority or brute force. Mastery over 'wilful beings', even in purely cooperative environments, involves the explicit exercise of power in all its guises. However, the greater the complexity of the social structure over which one is attempting to exert control, the more difficult that control comes to be.[66] Kosovo illustrates this well. In attempting to apply seemingly straightforward and universal values, the result was an intricate mish-mash of conflicting strategic priorities and rules of engagement (ROE) that ultimately placed greater operational constraints on the superior military force than on the outmatched Yugoslavs.

Primacy, risk, and dominance: the new operating system

To recall Mahbubani's appeal for a new contract between America and the world, the challenge of globalisation and the opportunities and hazards it presents to all states suggests that, while such a call may sound reasonable, it is unlikely to be heeded. Even when partners share close moral and social values, as within NATO, achieving common purpose in that most risky of international endeavours, military intervention, has largely proved elusive. Currently, no great power seeks to fundamentally alter the structure of the international system. Indeed, even former revolutionary powers such as China and Russia desire greater integration with international society, for instance through membership of the World Trade Organisation (WTO). The absence of competition has led, not only to a global failure to invest in military capabilities that would challenge American military predominance, but also an absence of capability that would permit alternatives to American-led military interventions.

Articulating power ultimately involves engaging in risky behaviour, whether through the development of science and technology or by applying force. The subjective nature of risk ensures that how it is defined, especially if it is a risk to values and norms as opposed to specific interests, will remain highly contentious. Just as it has been difficult to arrive at common definitions of justice and order in the present international environment, the intricate interplay of domestic forces will ensure that each state regards the risks it faces in highly contingent (and ultimately expedient) ways, frustrating common action. The contested and politicised nature of risk and the discursive formulation of collaborative power suggest that struggles between America and its partners during crises will be contests over how best to 'spin' the available information.

In reality, the international environment will often not wait for diplomatic conversations to play themselves out, and will present a succession of crises for states to deal with. The conjunction of globalised opportunities and hazards, strategic indecision, and US global agency through its military primacy prioritises American military operational methods. For the present, the command of the commons that the US enjoys ensures that only America has the capacity to act on a sustained, global basis. Thus, how the US defines risk will be the most important determinant for international action. Certainly, America's partners, and for that matter its adversaries too, will play important roles in the articulation of that risk. But because America's principal partners are all status quo powers,[67] they will seek to restrain it, rather than prod it into action, for fear of the possible strategic consequences for themselves.

For America, the key question may be how long to wait before acting. When it does move, the mission will decide the coalition. The Windows analogy that introduced this chapter is persuasive on a number of levels. In the current strategic environment, US military operational methods will structure the manner in which nations engage militarily, in the same way that Windows structures the global software environment. Local solutions, such as Israeli developments in urban operations or British ones in low-intensity warfare, are always possible. Where these are useful, America is likely to incorporate them into its own doctrine, just as Microsoft acquires smaller software companies to add value to its own suite of services.[68] However, nations seeking to participate in international military ventures will ultimately be forced to accommodate the American operational technique in the same way that software developers have had to come to grips with Windows. The next chapter discusses the nature of America's military operating system, and how it will interface with coalitions in this new environment.

2 Freedom and control

Networks in military environments

> As we prepare for the future, we must think differently and develop the kinds of forces and capabilities that can adapt quickly to new challenges and to unexpected circumstances. We must transform not only the capabilities at our disposal, but also the way we think, the way we train, the way we execute, and the way we fight.
>
> Donald Rumsfeld, 2003[1]

Claims to the establishment of revolutionary ideas are difficult to verify in the present: only the passage of time can truly confirm the impact an idea will have on history. For example, immediately after the Second World War, nuclear weapons were widely believed to have revolutionised war. Nevertheless, their role in warfare has to date been latent, rather than direct. Second, as Colin Gray points out, the concept of a 'revolution in military affairs' is essentially an interpretation placed on the unfolding of events, as opposed to an objectively verifiable occurrence with a time and place attached to it.[2]

Much is also unclear about current developments in military networks. As in the case of nuclear weapons, it may ultimately prove impossible to implement information technologies militarily in the manner predicted by NCW's early proponents. Moreover, non-US militaries may devise alternative approaches to the use of NCW,[3] just as the combined use of armoured forces, wireless communication, and aircraft took much trial and error by various powers before and during the Second World War.[4]

There is much that is promisingly novel about the military role that IT might play. But while this might, from some perspectives, warrant the label 'revolutionary', at the heart of NCW lies a basic dialectical tension. NCW promises faster, more precise, more decisive operations thanks to information-sharing. In this regard, NCW is oriented towards increasing the operational freedom of choice for military commanders such that they can avoid or efficiently surmount the barriers that war creates through the enemy's active resistance, as well as the ignorance that the danger and chaos of operations generate. At the same time, because military operations are ultimately undertaken to ensure the security of the state, the military context is an environment of strict control and direction. The lethality of warfare further accentuates the critical nature of this operational dimension. Information is too critical to be unregulated and the security of it is paramount. These two aspects – freedom and control, sharing and security – circle each other warily within the nature of NCW. If too much operational freedom is delegated to subordinate units, control is lost to commanders; if too much control is retained, operational flexibility is compromised.

Networks challenge the traditional hierarchical structure of military organisation; in the same manner, they also raise important questions regarding coalitions and how they will operate. Coalitions are centrally concerned with sharing – resources, influence, and information – and thus should be open to the use of networks. However, while a central premise of informationalism is the power generated through collaboration, networks can also be exclusive. US military primacy privileges America's own national secret-level network, the SIPRNET, over other nations' smaller ones. On the Internet, information has no borders; on military networks, however, it is absolutely essential that unbreakable frontiers are in place.

The origins of NCW

NCW is a relatively new concept, first appearing in the open literature in Cebrowski and Gartska's 1998 article, published in the US Naval Institute's journal *Proceedings*.[5] However, the idea of networking information amongst naval platforms began to emerge during the Second World War. The challenge presented to surface ships by aircraft, ubiquitous at sea for the first time with the appearance of modern aircraft carriers, required considerably more coordination amongst fighting platforms than had traditional naval gunnery.[6] The coordination of diverse vessels and missions resulted in the development of modern Combat Information Centres, or Operations Rooms.[7] Contemporary tactical data-exchange systems such as Link and the Global Command and Control System (GCCS) can also trace their origins to the Second World War.[8] Finally, cybernetic theory, which forms the basis for much thinking on information and control, was developed initially as an offshoot of ballistics research into the problems of anti-aircraft weaponry.[9]

After 1945, both the US Navy and Air Force continued to develop the role of information in the conduct of war; by the end of the 1970s, the US Army had joined them in this. Information has always been crucial to naval strategy, as navigation and locating the enemy are central to all naval battle. However, the US Navy's Maritime Strategy of the 1980s specifically exploited information-based technologies such as *Aegis* and advanced sonar to threaten the Soviet Union's coastline, thus potentially globalising any struggle over Western Europe.[10] Likewise, air and space technologies emerged at a steady pace after the Second World War, including advanced airborne radars and command and control systems, precision guided munitions, stealth aircraft, and satellite imaging.[11] Finally, the US Army's growing interest in operational warfare doctrines after the end of the Vietnam War led to concepts such as AirLand Battle. These required significant intelligence and the exchange of information between Army and Air Force units in order to coordinate deep strikes into Soviet rear areas.[12] After 1945 then, and with increasing intensity from the mid-1970s, each service independently pursued strategies with similar themes relating to the growing importance of information and its transmission and sharing. This serendipitous evolution was noticed by the Soviet armed forces in the 1970s, and the issue of an American 'military technical revolution' was discussed in their professional journals.[13]

In some respects, the close of the Cold War marked the end of a political era as well as a military one. The development of doctrines like the Maritime Strategy and AirLand Battle all pointed to the geographic expansion of the battlefield to something beyond what had been well understood, to that point, by 'operational art'. Operational art first appeared during the military changes of the early nineteenth century, when the enlargement of the battlefield, its growing complexity due to the rapid introduction of new technologies, and the growing role of the state's economic power in fielding and sustaining military forces led both to longer military campaigns and to theatre-scale warfare.[14] Aside from a solid grounding in tactics, successful military commanders needed to come to terms with the time and space dimensions of moving numerous large and complex military formations to achieve the ends of strategy. In the eyes of many strategic analysts, operational art reached its acme at the end of the First World War.[15] Former British Brigadier and historian of the First World War Jonathon Bailey makes the bold assertion that:

> Three-dimensional conflict was so revolutionary that the tumultuous development of armor and air power in 1939–45 and the advent of the information age in the decades that followed amounted to no more than complementary and incremental improvements upon the conceptual model laid down in 1917–1918.[16]

The operations projected by the US military at the close of the Cold War were inherently global in nature, however. The ability to deal with the complexity of this battlefield was greater than the individual competency of any single service, a point recognised by the introduction of the terms 'battlespace' and 'war-fighter' in the 1990s.[17] Just as business was dealing with the challenges of an enlarging global market by exploiting IT, so too the US armed forces were dealing with operational challenges on a similar scale, and exploiting the same sort of technology. By the mid-1990s, the US military was putting these new developments into doctrinal perspective.

The emergence of the concept

In 1996, Admiral William A. Owens published his article 'The Emerging System of Systems' in the National Defense University's journal *Strategic Forum*. This described a concatenation of sensors, command and control systems, and precision weaponry that would, he argued, result in 'dominant battlespace knowledge'.[18] In the same year, *Joint Vision 2010* appeared, describing the 'conceptual template … for achieving dominance across the range of military operations through the application of new operational concepts'. *JV2010* introduced the concepts of Dominant Manoeuvre, Precision Engagement, Focused Logistics, and Full Spectrum Protection to achieve 'massed effects'. *JV2010* represented the distillation of 20 years of technological advance and operationally focused thinking in the US armed forces. Yet it was clear that 'information superiority' was the basis for these novel operational concepts. To that extent, they amounted essentially to a more elaborate restatement of the 1980s-era AirLand Battle ideas. *JV2010* incorporated conceptual advances in manoeuvre and joint warfare, but operations were fundamentally derivative of what had preceded before. While *JV2010* spoke of the emergence of the Revolution in Military Affairs (RMA), a further step was required before one could begin to call these developments truly revolutionary.

The elaboration of NCW

Following Cebrowski and Gartska's seminal article, NCW was elaborated in three semi-official publications: *Network Centric Warfare*, written jointly by Gartska, Director of Research and Strategic Planning for the Office of the Undersecretary of Defense (C3I) David S. Alberts, and retired US Army Colonel Frederick P. Stein, published in 1999; *Understanding Information Age Warfare*, by Alberts, Gartska, Richard E. Hayes, and David A. Signori, published in 2001; and *Power to the Edge: Command and Control in the Information Age*, by Alberts and Hayes, which was published in 2003.[19] Together, these three works form the canon from which most thinking on NCW has developed. Through a series of business case studies, *Network Centric Warfare* introduces the idea that networks generate power through the distribution of information. *Understanding Information Age Warfare* takes the idea of NCW and develops a theory about how information, knowledge, and awareness interact in a military environment. *Power to the Edge*, a more conceptual piece, ruminates on the implications of information and networks for military organisations and their operations.

In exploring how computer networks are altering the economic and business activities of US corporations, *Network Centric Warfare* shows its descent from earlier works by Alvin and Heidi Toffler, who suggested in their influential *War and Anti-war* that 'the way we make wealth is … the way we make war'.[20] Corporations, having linked together 'knowledgeable entities' (sub-units within the organisation) through computer networks, can take advantage of the shared awareness thus generated to make decisions faster and more efficiently, and to improve the accuracy of business predictions. Networked businesses may also improve collaboration between sub-units, and may ultimately create efficiencies in their supply chains and customer relations. *Network Centric Warfare* suggested that the compression of time and space caused by this shift would also impact on warfare. In essence, the same processes so important to creating better business decisions would also enable military commanders to create a condition of 'information superiority', analogous to earlier concepts of air superiority or sea control.[21]

Such capabilities would be increasingly important because of the growing complexity of the modern battlefield.[22] This new approach would produce a series of remarkable outcomes changing the very nature of warfare. Networks would permit the generation of combat power from highly dispersed yet agile military units because of their enhanced situational awareness. The authors of *Network Centric Warfare* argued that both the 'fog of war' and friction in military operations, while not eliminated completely, would be dramatically reduced.[23] As enhanced awareness would reduce risk, the cost of operations would decline, just as networks permitted businesses to reduce their costs.[24] The combination of these assets would permit networked militaries to create 'mass effects', instead of massing forces.[25]

In *Understanding Information Age Warfare*, these ideas are fleshed out into a full theory of operations. The authors begin with a series of assumptions about how experience ultimately translates into awareness, from which they derive a theory of warfare in networked environments. They suggest that we should consider the manner in which we obtain information about the external environment through the interaction of a set of logical assumptions. Sensory impressions of the environment can be directly experienced (seeing an event occur, for example) or indirectly inferred (through the interpretation of data from a sensor such as a radar). These impressions are then translated into 'information' by putting them into a 'meaningful social context' by identifying patterns through a comparison made between the sensed data and what is already known about the environment. These patterns represent 'knowledge', and comparisons between what is 'known' about the world (prior knowledge) and what is currently being sensed generates 'awareness'. Finally, with sufficient levels of knowledge, by identifying developing patterns the observer can draw inferences about what is likely to happen. In this way, awareness permits the observer to identify what is known about the past and present, while 'understanding' allows identification of 'what the situation is becoming'. At the end of this sensing process, the observer is capable of deciding what to do, and then acting on that decision. The whole process is similar to the famous 'OODA' (Observe, Orient, Decide, Act) loop developed by Colonel John Boyd.[26]

Alberts and his colleagues describe the world in which this process of sensing and interpreting data takes place as a series of interconnected 'domains'. Three principal ones are posited. The 'Physical Domain' is described as the scene where all action takes place. It is the location where military forces manoeuvre, strike, and defend themselves, and action, being directly observable here, can be measured through direct and indirect sensing. The 'Information Domain' is where information is created, manipulated, and shared. It is a virtual environment in which data are transferred and shared amongst actors through technology, and software; at its heart, it is a medium for communication. The 'Cognitive Domain' resides in the minds of the actors participating in the network. In this domain, understanding is created through the interpretation of the data being communicated from the physical domain through the information domain. It is in the cognitive domain that information is evaluated and judged, and decisions made.[27] To these three domains, *Power to the Edge* adds a fourth, the social domain, which mediates the evaluations, judgements, and decisions developed in the cognitive domain.

As Alberts and his co-authors point out, NCW is principally about sharing information and awareness.[28] It thus enables the development of superior awareness that ultimately translates into information superiority. This is described as the 'NCW Value Chain', which was first elaborated in *Network Centric Warfare* (as shown in Figure 2.1). This figure describes the series of inferences that lead ultimately to the establishment of increased combat power. By lowering the costs and risks associated with military operations, greater effects can be generated. Essentially, then, as the 'Tenets of Network Centric Warfare' assert:

> a robustly networked force improves information sharing and collaboration, which enhances the quality of information and shared situational awareness. This enables further collaboration and self-synchronisation and improves sustainability and speed of command, which ultimately result in dramatically increased mission effectiveness.[29]

Shared knowledge is critical for forces participating in a networked operation.[30] The end result of this sharing of information and awareness is the creation of additional combat power through enhancing the utility of information provided to decision makers. Information can be characterised by its richness (or its quality)

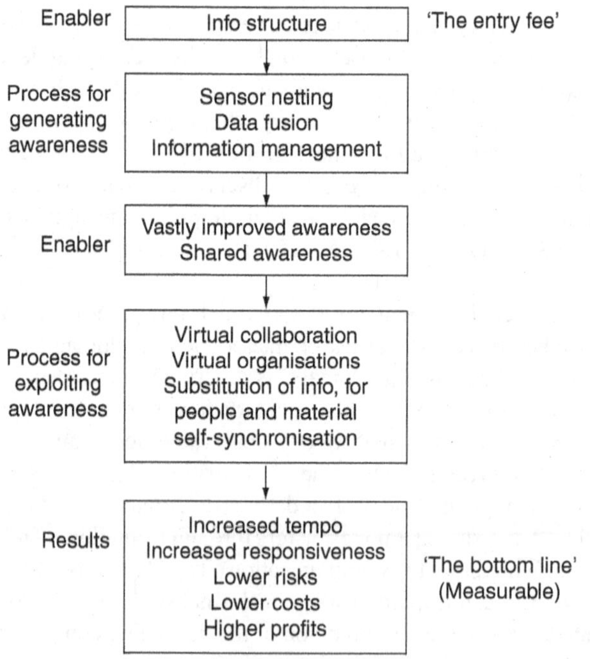

Figure 2.1 Tenets of NCW.

and its reach (or its ability to permeate every area on the network). Typically, the richer the information, the less reach it has. This is most obviously the case with classified information, which is generally closely held by those with a 'need to know'. However, those in the field with proper clearances may be unable to access this information because of their distance from those who control it. Lower level information will spread much further through a network than the most highly classified material.

In a functioning network centric environment, however, richness no longer faces barriers to its reach. Those with the proper credentials in the field will be able to access even highly classified information in real time, thereby generating additional combat power.[31] A 'common operating picture' permits greater unity of command and purpose and de-conflicted missions, avoids duplication of effort, enhances early warning (and thus greater force protection), and allows resources to be used more economically.[32]

The requirements are, however, high. In the physical domain, all elements of a military force must be connected together, 'achieving secure and seamless connectivity and interoperability'. In the information domain, people and platforms must be able to access, share and, most importantly, *protect* information 'to a degree that [they] can establish and maintain an information advantage over an adversary'. Finally, in the cognitive domain, forces must be able to use this shared information to develop awareness of their environment, and share that awareness with other network participants. Unless these objectives are accomplished, military forces will be unable to 'self-synchronise', and thereby take advantage of the benefits conferred by the network.[33]

While it is the combined effect of the four domains that allows shared awareness and self-synchronisation, the lynchpin of the whole enterprise is the *security* of the information domain. Establishment of a combat advantage depends on information superiority, but this superiority must be protected. In the words of Alberts *et al.* 'in the all-important battle for information superiority, the information domain is ground zero'.[34]

With a theory in place describing the relationships between information, knowledge, and awareness, further thinking concerned the implications for military operations in this new environment. The conclusions of this research emerged in 2003 in *Power to the Edge*. Here, Alberts and Hayes argued that, in order to take advantage of the opportunities offered by NCW, militaries would have to 'focus on C2, where information is translated into actionable knowledge'. In the modern battlespace, traditional procedures and organisations for the command and control of military forces would be unable to cope with the complexity that these forces will face. Alberts and Hayes argued that militaries had so far been able to adapt by using 'work-around' procedures that were typically unique to the time and place of a specific operation. Relying on these inefficient information-sharing practices in the face of the growing complexity of the modern battlespace will eventually frustrate the application of military power. Decision makers in these challenging global arenas cannot possibly anticipate every outcome, nor do they possess complete knowledge about the environment in which they will operate. In order to maximise the potential offered by information, modern organisations must be capable of sharing their specific situational awareness with others.[35] Furthermore, since they cannot know who they will work with, nor which systems may be relevant, a high degree of agility would be necessary 'in terms of who participates as well as who plays what roles'.[36]

Given these observations on the demands of the modern military environment, the centralisation of command and control is increasingly impractical. Instead, power needs to be devolved to 'edge entities':

> Power to the Edge involves the empowerment of individuals at the edge of an organisation (where an organisation interacts with its operating environment to have an impact or effect in that environment) or, in the case of systems, edge devices. Empowerment involves expanding access to information and the elimination of unnecessary constraints.[37]

This vision is potentially revolutionary: in terms of its organisational and procedural implications, it strikes directly at the hierarchical structures that militaries have always relied on for command and control. It remains to be seen whether militaries will be capable of adapting to such a wide-ranging vision. Nevertheless, to illustrate the Pentagon's commitment to it, Albert and Hayes point to the development of the Global Information Grid (GIG), which will integrate communications and computer systems into a secure, seamless 'infostructure providing access to a variety of information sources and information management resources'.[38]

The emergence of the GIG: networks and global military operations

The introduction of the GIG as a fundamental structural component of America's defences[39] points to the role of information technologies in transforming modern societies. Comparisons are easy to make between the military GIG and the civilian Internet. Transformation itself seems to be guided by an 'Internet paradigm' in terms of its overall vision.[40] In testimony before the US House of Representatives Armed Services Committee in 2004, the Assistant Secretary of Defense for Networks and Information Integration, John Stenbit, described the GIG as a 'private World Wide Web' that would 'support the transformation of our warfighting and business practices'.[41]

Under current plans, the GIG will establish its core capabilities by 2010, at a cost of $21 billion. However, full implementation is not expected until 2020. By then, the GIG will 'integrate all [Department of Defense] information systems, service applications, and data into one seamless and reliable network'.[42] Structurally, the GIG will be realised through four related endeavours: the Global Information Grid Bandwidth Expansion (GIG-BE),[43] the Transformation Communications System (TCS),[44] Network Centric Enterprise Services (NCES),[45] and the Cryptological Transformational Initiative (CTI).

Some analysts have speculated that such 'super networks' are inevitable.[46] Several features of modern operations contribute to this impression. The steady expansion of the operational battlespace since the eighteenth century and the globalisation of American defence tasks have demanded greater coordination amongst armed services. Missions such as close air support, the suppression of enemy air defences, missile defence, and deep-strike operations all require close coordination amongst highly disparate force elements, many of them crossing service boundaries, some traversing traditional theatre and command boundaries. In order to accomplish missions such as these, even de-confliction of effort requires a high degree of communication and coordination between participating units. To go the next step and ensure effective joint coordination demands highly integrated planning.[47] Moreover, the human, economic, political, and social costs of sending US forces abroad are increasing. Moving information instead of troops allows administrative, logistical, intelligence, and other support to remain in the US, even during periods of combat.[48] Even the force providers themselves may not need to deploy in the massive fashion of traditional combat operations, relying on the speed, agility, and manoeuvrability brought about by rapid and ubiquitous information-sharing.[49] Finally, the importance of building stability in regions of civil war and social breakdown has helped to generate the complex battlefield identified by former Commandant of the Marine Corps General Victor Krulak as the 'Three Block War'.[50]

The highly complex nature of these shifts in the military environment prevents any one commander or organisation from possessing complete awareness of all critical aspects affecting operations. Geographically dispersed and organisationally complex operations, which may involve relatively small forces by traditional standards, require extensive information-sharing. Networks assist in the planning and conduct of operations in many ways. Instantaneous communications have both shrunk the world, and accelerated decision making.[51]

In this complex environment for the US armed forces, the very malleability of networks is also attractive. Information age sociologist Manuel Castells has pointed out that 'nodes' on a network vary in terms of their overall relevance. The importance of any given node on the network stems not from its function or features, but from its ability to contribute to the goals established by the network. Nodes can be added or deleted from network architectures as their importance changes, or as the missions alter. This permits considerable flexibility (in determining the paths along which information can be sent), scalability (in terms of the growth or contraction of the architecture), and survivability,[52] allowing easy access to information 'anytime, anyplace, with attendant security'.[53] If fully realised,

> perhaps the single most transformational and operationally significant attribute presented by the GIG vision will be that US servicemen and women 'at the edge' will no longer be at the mercy of someone remote from the fight determining what information they need.[54]

Just as information superiority formed the base on which *JV2010*'s advanced concepts rested, *information sharing* forms the base on which the edifice of military transformation rests.[55] Information sharing is often conflated with intelligence sharing. Intelligence sharing is an important aspect of information sharing; however, it is only a subset of it. Information includes not just intelligence, but also sensor information, planning information, and situational awareness. The 'fog of war' is commonly blamed for the waste of lives and resources associated with battle, and the failure of forces to achieve their purposes; the authors of *Network Centric Warfare* assert that any such fog is largely caused by a lack of battlespace awareness stemming from the inadequate distribution of information. Confusion stems from 'our inability to tap into our collective knowledge or the ability to assemble existing information, reconcile differences, and construct a common picture'.[56] While *Transformation Planning Guidance* blandly defines transformation in highly general terms,[57] the transformation necessary to overcome the fog of war and achieve the vision portrayed above revolves around 'seamless' information sharing.[58] As *Network Centric Warfare* points out, information superiority is 'in part gained by information operations that protect our ability to collect, process, and disseminate an uninterrupted flow of information while exploiting and/or denying an adversary the ability to do the same'.[59] Ultimately, the ability to build a *collective* awareness upon the *collected* and limited awareness of platforms and individuals operating in the battlespace constitutes the basis of America's military transformation plans.[60] In the words of one USAF officer '(Internet Protocol) brings global connectivity to the kill chain'.[61]

Information vulnerabilities

This powerful vision for warfighting contains within it a significant vulnerability. The same technology that enables dispersed and small formations to magnify their operational power through information sharing also enables an adversary to both read the intentions and plans of a military force, and alter the information to accomplish a variety of ends. The problems of unauthorised access to information sites that are supposed to be confidential is so widely understood as to have infiltrated popular culture; similarly, we are increasingly familiar with the threat posed by identity fraud, if not specifically in terms of national security. The threats of information denial and the clandestine alteration of stored data are less commonly appreciated, though just as damaging.[62] With the exception of a 'denial of service' attack, all of these methods involve penetrations of secure systems. Identity fraud is the digital equivalent of introducing a 'mole' into a supposedly secure organisation. The damage that 'malicious insiders' can cause to information systems points to a fundamental change in the nature of warfare. Traditionally, defence has always been the stronger form of warfare, but relations between offence and defence are reversed in terms of information security. As a study by the National Academy of Sciences described it:

> Imagine a situation in which truck bombers in a red truck attempt entry to a military base. The bomb is discovered and they are turned away at the front gate, but allowed to go away in peace to refine their attack. They return later that day with a bomb in a yellow truck, are again turned away and again go away in peace to refine their attack. They return still later with a stolen military truck. This time the bomb is undetected, they penetrate the defenses and they succeed in their attack. A base commander taking this approach to security would be justly criticised and held accountable for the penetration.[63]

The difficulty of establishing identity in a digital environment[64] highlights the danger such penetrations pose to the security and integrity of an information environment.

A second challenge to information security on the GIG comes from authorised users of the system, who might compromise information from simple ignorance. In its essence, digital information is persistent and transportable: it is easily copied, archived, and shared. The implications for inadvertent disclosure and subsequent propagation of classified information are evident. The Google search engine routinely archives all information it categorises, permitting users to view material that has since disappeared from the web pages on which it was originally placed. The same miniaturisation developments that have enabled electronic communications have also eased the problem of transporting large amounts of data over distance. Networks permit the rapid replication and translocation of information in ways that, in the past, spies could only dream of.[65]

Control versus anarchy: the problem of information assurance

These essential issues of information vulnerability have not gone unnoticed by US security agencies. Nevertheless, there has been no fundamental progress on information assurance, in contrast to the rapid developments in communication links and information sharing since the 1990s. In the 1990s, according to the US Government Accounting Office (GAO), the Defense Information Assurance Program, although it made limited progress, ultimately failed to meet its goals.[66] In 2004, the GAO identified three issues posing particular challenges for the GIG: deciding when and how much information should be posted; establishing rules to ensure that the GIG could work securely without compromising the benefits of flexible and dynamic information sharing; and convincing data owners of the value of sharing data with a broader audience and trusting the network sufficiently to post it. All three point to the critical role played by information security.[67]

The GIG's development programme subsumes the Cryptological Transformation Initiative, a $4.8 billion project funded by the National Security Agency and involving the development of advanced firewalls, multilevel security protection, and High-Assurance IP encryptors.[68] Any information assurance system, however, has to accomplish a variety of goals. As defined by the US Department of Defense, information assurance is:

> measures that protect, and defend information and information systems by ensuring their availability, integrity, authentication, confidentiality, and non-repudiation [which] … includes providing for the restoration of information systems by incorporating protection, detection, and reaction capabilities.[69]

In accomplishing these tasks, however, the systems engineer confronts the essential nature of digital information. Examining the sub-concepts embedded within the above definition of information assurance, '*availability*' of information is for '*authorised*' users only; '*integrity*' of information means protection from '*unauthorised*' change; '*authentication*' involves verifying the *identity* of the originator of a request for/author of data; '*confidentiality*' involves ensuring data from '*unauthorised*' disclosure; and finally, '*non-repudiation*' involves incontrovertible proof of the *identities* of those using information on a network.[70] Each of these aspects relates either to the fundamental authority, or control, over how and whether network data are to be stored, shared, and manipulated, or to the identification of participants using such data. In sum, the challenge of ensuring security casts a large shadow over digital collaboration in military environments. Control of information in terms of both its security and its proper interpretation is of paramount importance.[71]

These requirements[72] suggest a fundamentally different orientation to information compared with the events that led to the emergence of the Internet. Indeed, the control necessary to guard against even a single point of failure in information security suggests elements of a police state where 'every node is a sensor that can relate security information to those tasked with securing the network'.[73] This vision of near-totalitarian control of information clashes with the anarchic nature of the Internet itself.

In contrast to the highly controlled military networks, no single authority controls the Internet. It is an 'anarchical society' in the same manner that Hedley Bull described the international environment.[74] Just as in international relations, the absence of authority does not preclude a degree of order: the technical protocols for the transmission and sharing of information (TCP/IP and HTML and its variants) for example. While legal regimes are being established, in a global communications environment they depend largely on self-interested enforcement and compliance in a manner similar to international law. Still, there is enough order to permit the global conduct of a considerable amount of industry and business in this anarchical electronic environment.

In the same manner that concepts of justice are internationally contested, the Internet is also anarchical in terms of how truth is understood. There are no gatekeeping features on the Internet.[75] The introduction of web sites like Wikipedia take advantage of this aspect as well as the dynamic, malleable nature of digital information.[76] The popularity of 'blogs' and their growing influence on news reporting within the media is a similar issue. The ability of sites like the 'Drudge Report' to unearth key political scandals in Washington is due to the fact of differing approaches to how truth is mediated between blogs and traditional news organisations. The commitment to professional standards of reporting by mainstream publications like the *Washington Post* ensured that rumours of an affair between White House interns and the President were unpublishable without iron clad sources. The same aspect leads to the widespread academic criticism of Wikipedia.[77] In effect, the Web becomes a location for debate over truth owing to the multiplicity of sites presenting differing slices of reality, permitting web-surfers to arrive at their own unique conclusions. The effect of this is of course the development of sub-communities convinced of 9/11 conspiracies, or American possession of UFO technology at its base in Groom Lake ('Area 51').[78] Irrespective of these charges, the interpretation of truth on the Internet is similar to how abstract terms such as justice and freedom are interpreted in the international environment: each of these problematics owe their origin to the anarchical setting in which they are situated.[79]

That the Internet should display these anarchical features is not entirely surprising, according to Castells. In his analysis of what he calls the 'Network Society', its emergence was influenced by three key features, including the culture of individual freedom inculcated in both America campus environments and the counter-culture movements of the 1960s. The peace movement, civil rights struggle, and growth of environmentalism during this period were founded on the defence of civil liberties, the advancement of free speech, and the opposition to traditional sources of authority. Similarly, the academic culture of universities, especially those in the United States, was that of shared discovery in which interpersonal professional communication was the basis for academic progress and the advancement of truth. Each of these movements 'stood in sharp contrast to the world of corporations and governmental bureaucracies that had made secrecy and intellectual property rights the source of power and wealth'.[80]

Pekka Himanen asserts that the information sharing on which the 'network society' and its electronic sinews are based has permitted the establishment of a 'culture of innovation', sometimes referred to controversially as the 'hacker ethic'.[81] The spirit of this culture is one of innovation, individuality, and networking. It approaches work as a child does play, and emphasises the value of creation over the spirit of the profit motive. 'Money centredness leads to the closing off of information. Innovation lives on the open flow of information'.[82] This orientation towards information, freedom, and innovation has also inspired technological movements such as the Open Source Initiative and the associated developments of the Linux operating system.[83] The free exchange of information has undoubtedly formed the basis of the explosion of technological and scientific advances of the late twentieth century. Whether militaries can take advantage of the innovation that stems from such sharing will be very much dependent on their willingness to compromise strict information security protocols that impede such sharing.

The fundamental dialectical tension within the network centric vision

This necessarily limited discussion of the role of information and networks in modern military thinking, and the development of the GIG emerging from the nature of NCW as contrasted with the development of the Internet and its impact on modern society, suggests the tensions that underlie these developments. On the one hand, we can note that the role information exchange has played within military contexts is an aspect of warfare that has a long history and hardly constitutes a revolutionary development. What does seem to be revolutionary is the near instantaneous information sharing on a global basis due to developments in IT. The potential offered by these technological developments seem to suggest new approaches to both how time and space function in military operations, and reflects changes in terms of fundamental principles such as that of mass and concentration.

The power that militaries may derive from networks comes at the price of ensuring the security of the information domain from direct attack or clandestine infiltration. The complexity of this mission in a digital environment, where concrete identities are difficult to establish, suggests the need for a level of control over information that contrasts starkly with the nature of networks in civilian society. In effect, the military use of IT seeks to exploit their capacity for innovation, creativity, and the expansion of knowledge. At the same time, however, networked militaries require a level of control to protect the operational advantages networks give them.

This risk may be somewhat exaggerated.[84] The Open Source Movement itself accepts the positive role that trade secrets play in some product development where competitive advantage is generated through research and development that must be protected against competitors.[85] Similarly, an examination of the principals that underlie the Open Source Movement and, indeed, Hinamen's culture of innovation itself call to mind the military principle of *auftragstaktik*.[86] Control and restraint have featured strongly in the architecture of the Internet. Control through information is central to the notion of cybernetics,[87] and there is plenty of literature[88] on the Orwellian nature of databases and the centralised control of information in government and society. Nor do digital insurgents necessarily enjoy greater technological advantages over GIG users due to their ability to manipulate data freely within the anarchical environment of the Internet: the dialectic between innovation and control will play itself out in both the civil and military domains of the network environment, and will do so in unpredictable ways.

Networks enhance power through their scalability, survivability, and flexibility. The ability of actors to take advantage of these features, however, depends on 'the pattern of power present in the [structure] of the network'.[89] As Cebrowski observed, if 'you are not on the net … you are not in a position to derive power from the information age'.[90] But just as not everyone on the planet is able to access the Internet, so not all states' military forces are able to interoperate effectively with those of the United States. In many studies, this has been blamed on inadequate capital investment in IT, or on a failure of US technological developments to facilitate high levels of allied interoperability. This focus on a growing 'digital divide' misses the point that fragmentation is a structural feature of networks; as Castells puts it, networks 'search for valuable additions everywhere in order to incorporate them, while bypassing and excluding those territories, activities, and people that have little or no value for the performance of the tasks assigned to the network.'[91] This results in a differentiation between those who are sources of innovation, those who simply carry out instructions, and those who are irrelevant as either workers or consumers.[92]

The role that coalition partners will play in shaping the larger network and the goals towards which it will work will not necessarily be determined by technical capability or the ability to 'interoperate'. While 'plug-and-play' interoperability will undoubtedly be important for coalition partners, who are allowed within the larger confines of the network, and the role that they play once they are there, will be determined by the political value they bring to any endeavour, as established by very traditional national interests. In most coalition operations, the SIPRNET will be the most important network, and so America will establish the policies under which data will be shared between coalition partners. Questions must be posed as to whether coalition partners can also play roles as innovators within a network, or whether they will be relegated to less powerful roles, drones in other words. The term 'flags around the table' heard frequently in the context of coalition operations, neatly captures the reality of partners as politically valuable but militarily irrelevant players. And so we return to the issues of freedom and control. For all the latitude digital information provides, a strict logic constrains its users. Even as it offers greater operational freedom for military commanders, information assurance limits complete freedom of action. Even in the most liberal of national security networks, information assurance guarantees that coalition interoperability will be subject to an extraordinarily high degree of control. If unfettered trust is difficult to establish in a purely national setting,[93] then its achievement in a multinational military network is most unlikely, even between the closest of allies.

3 International anarchy and military cooperation

While capable of mounting large, rapid, and decisive operations to achieve certain limited ends, the United States remains tied to its allies to assist it in maintaining international order: as technologically advanced and well funded as the US military is, it simply lacks the manpower to be in all places at all times.[1] At the same time, however, the march of technology presents growing difficulties to America's allies and partners in their ability to support this goal. The US is not ignorant of these challenges. The American military, particularly by its Joint Forces Command in Norfolk, VA, has devoted a great deal of effort to studying the problem of coalition interoperability. The change in name in 2005 of the annual Joint Warfare Interoperability Demonstration, which highlights new technology that seeks to improve interoperability, to the Coalition Warfare Interoperability Demonstration (CWID) highlighted the seriousness with which the Pentagon views the issue. Nevertheless, the evolution of military technology and doctrine towards information-centric models presents issues that will be difficult to resolve. The problem stems from the evolution of NCW as an operational concept devised to create a competitive edge by taking advantage of technological developments at the tactical level that is now driving larger strategic issues in terms of military cooperation between the US and its partners.

Political constraints, imposed by the nature of the international environment and the role played by warfare within it, will ultimately frustrate the technological projects of the US to keep its alliance partners fully engaged in its security policy. The ultimate result of this development may be increasing unilateralism in US security policy, and a growing reliance on America's partners to play subordinate roles, limited largely to peace support, stability, and reconstruction operations. The free flow of information sought by NCW will be critically limited by the impact of international anarchy on state-to-state cooperation. The fact that states have different interests will further hamper cooperation in security endeavours, especially during this period of limited warfare that characterises the unipolar international environment. Thus, the current global political environment will obstruct the drive to increase levels of interoperability between militaries. As such, prognostications that NCW will represent a new paradigm for military operations will ultimately prove hollow for those who seek to operate with the US, including its most trusted military allies.

It is well recognised that NCW is changing how militaries operate, both in battle and in 'operations other than war', and that the sharing of information can only grow in importance as armed forces continue their never-ending quest for competitive advantage. It is also axiomatic that the potential for failure in coalition operations exists should partners diverge too greatly in terms of their ability to operate together. There remains hope, however, that technical means may obviate this problem. The search for an 'interoperability black box' continues to attract the United States and its closest allies, which have established various forums to explore this issue. These concerns are being addressed by the ABCA (Australia, Britain, Canada, and America) nations. Seven countries[2] have also established the Multinational Interoperability Council (MIC) to explore common concerns. The CWID is also part of this effort. Finally, limited operational experiments have been developed by the J9 organisation of Joint Forces Command and its allied partners, to test evolving concepts of information exchange in operational scenarios.

However, actors in both the technical and policy communities have not yet recognised that the problem they are attempting to resolve may have no answer. Policy barriers hinder allies, even close allies, from sharing information with each other transparently and, consequently, issues of allied and coalition interoperability will not be easily addressed. Indeed, as IT becomes ever-more important to military operations, the United States may over time find itself able to operate with fewer and fewer partners.

The international environment and military cooperation

How the nature of the international environment and the role states play there will affect the assumptions made by NCW theory about international military cooperation is an unexplored question. The principal reason for this is that explorations of networking have concentrated on the technological issues involved. Moreover, the literature is dominated by American authors writing about American developments and experiences. Allied perspectives have been limited largely to major partners, notably the British, who still have some capacity for independent action and unilateral operations.[3]

As Snyder points out, 'anarchy is the basic cause of alliances and their Achilles heel'.[4] Security fears create the need for alliances, and yet the same anarchic conditions lead to doubts as to the reliability of any agreement made at the international level. Shared interests and threat perceptions create mutually dependent interstate relationships, but do not necessarily ensure perfect cooperation between partners, even within alliances. As a state can never be certain that its partners will completely fulfil their obligations, each state participates with the goal of minimising its contribution while maximising its partners' obligations.[5]

The distinguishing characteristic between alliances and coalitions concerns the extent of the shared interests that bring them into being. While coalition partners may share some interests, they do not do so to the same depth or for the same length of time as true alliance partners. Indeed, coalition partners may be competitors in other areas, or may choose to oppose each other even on related issues. A coalition organised around a particular issue thus tends to be driven by the state that possesses the greatest interest in that issue. The level of cooperation it can expect will vary in line with how closely its interests match those of its partners.

In both alliances and coalitions, bargaining over strategic direction and operational tasks assumes the form of placing the alliance or coalition deliberately at risk in order to coerce partners into acquiescence. In military cooperation, the bargaining power of any state is related to its overall dependency on its alliance partners. States that are able to effectively exploit asymmetrical relationships within an alliance will gain greater bargaining power.[6] As Snyder puts it, 'dependency refers to the degree of harm that partners could inflict on each other by terminating the relationship…'.[7] Those with a crucial supply of an important asset, be it military or diplomatic resources, will enjoy enhanced bargaining power with their partners. By threatening to deny access to those resources, they are able to manipulate their partners' fear of abandonment to extract concessions from them.[8]

The military dependency of one state on another is powerfully shaped by international structure through the medium of threat perception. Polarity effectively determines the rigidity of military obligations each state owes to its partners. Multipolar environments are marked by high degrees of fluidity between partners, whereas bipolar environments are far more stable.[9] The present unipolar environment has had a distinct impact on how states approach security cooperation. Contemporary coalitions are characterised by the speed of their formation, their tendency to coalesce around issues of peace support and international stability, their lack of a strict hierarchy (and thus the absence of any disciplinary features), and the relative lack of strong national interests guiding their creation (which means that the cost of withdrawing from them is relatively small).[10] This is to be expected when a single state represents the sole guarantor of international order; there is general agreement that existing norms represent all states' best interests, and no competing power is available to impose substantially different ones. As such, in the post-Cold War period, we have tended to see far more flexible and temporary 'coalitions of the willing', including among states with formal alliance relationships, as with NATO and the United States in Afghanistan.

Limited war and interoperability

Kenneth Gause of the Center for Naval Analysis is one of the few writers on NCW to recognise that interoperability is not just a question of technology, but one that also concerns the nature of participation:

> For those allies that want to operate closely with the US in prominent positions, even in high threat environments, the level of interoperability will have to be high, possibly bordering on seamless. However, for other allies, the demands of interoperability will be lower.[11]

Still, Gause makes no mention of the role politics may play in shaping decisions on the level of participation. How directly a state is willing to commit itself to any given conflict will have a direct impact on the level of interoperability between partners at all levels of warfare.

Commitments to fight in a particular war or decisions to align with a specific nation are based on strategic rather than operational rationales; alliances made during war differ significantly from those entered into during peacetime. Wartime pacts are made against a particular country or countries, while peacetime pacts are usually less specific.[12] In war, alliances are frequently temporary and are aimed at winning. They are general in nature and comprise the total interests of the parties. In peacetime, however, alliances are more commonly limited to a fraction of the total interests of a state.[13]

Limitations in warfare are generally described in terms of geography (where operations can or cannot take place), objectives (how victory is defined), means (what weapons will be employed), and targeting (whether to engage in counter-force or counter-value strategies).[14] Clausewitz remarked that war naturally tends towards the maximum effort if left unchecked. However, states will not commit forces blindly to a conflict, but instead invest according to the objectives that are sought. As the danger to national survival increases, so too does the willingness to subordinate self-interest to the overall collective effort.[15] In military cooperative ventures, Clausewitz argues that:

> The question is then whether each state is pursuing an independent interest and has its own independent means of doing so, or whether the interests and forces of most of the allies are subordinate to those of the leader. The more this is the case, the easier will it be to regard all our opponents as a single entity, hence all the easier to concentrate our principal enterprise into one great blow.[16]

If only allies were mercenaries, then the issue of what they would be willing to do in order to achieve the war's objectives would be moot. The 'extreme danger', to use Clausewitz's term, represented by Hitler or the Soviet Union forced a level of cooperation between Western states that was in many ways unprecedented. The slow collapse of the communist threat to the West and the ultimate disappearance of the Soviet Union has since set up reverberations within the Western alliance that have yet to resolve themselves. Still, it is readily apparent that the calculation of interest in committing to new political objectives has become more and more blatant within NATO.

The emergence of this 'natural' alliance behaviour will be apparent even in American actions. The original US mantra during the Cold War was most clearly spelled out by the Kennedy administration: America was willing to go anywhere and pay any price. It was the doctrine of automatic, reflexive commitment, of 'strategic coupling' and assured destruction. But in the current unipolar environment, America has moved more cautiously, only reluctantly involving itself in commitments in the Balkans and Africa, or being dragged into conflict by the pace of events, as in the Middle East or Central Asia. Vital interests – 'interests that are worth supporting militarily at the cost that must be paid'[17] – can no longer be simply taken as a given, but must be calculated anew for each confrontation.

Similarly, the interests of America's traditional allies cannot be taken for granted as they often could be during the Cold War. Maintaining Alliance cohesion has become remarkably more difficult in the post-Cold War environment, with each new American overseas engagement provoking greater questioning from allies. With each new commitment, the ability of the Western alliance to speak with a single voice has declined, and with it NATO's ability to deter its adversaries. This was readily apparent in the wrangling that occurred over Kosovo.[18]

Indeed, who is defined as an 'adversary' has itself become increasingly controversial. In coalitions and alliances, because of different interpretations of the problem or threat and the uncertainty surrounding allied reliability, these issues have become highly politicised. Differing assessments of risk mean that the very conduct of operations has become charged with political significance, rather than being conducted in the most efficient fashion possible. In other words, in cooperative military endeavours, unless it is an issue of pure and immediate survival, politics will always trump strict military necessity.

Unipolarity, NCW, and the possibility of seamless interoperability

In many ways, the United States has been successful in finding 'work-around solutions' to the problems of connectivity. While there were significant interoperability problems in the Balkans, some were resolved through the installation of American technology in allied formations.[19] Similarly, the US often devises procedural work-arounds in order to facilitate greater allied cooperation. This has been most evident in the Canadian integration into American carrier battle groups in the Persian Gulf throughout the 1990s,[20] and in the coalition naval operations of the War on Terror in the same region since 2001 (discussed in the next chapter). There would seem to be a limit on how far the United States is able or willing to go in attempting to solve some of these connectivity issues, however. This limit is defined first by the demands for information security, and second by the nature of trust between partners.

The search for greater operational freedom is the principle that animates the quest for information access under the NCW concept. In theory, universal access to common databases will lead to shared awareness and thus the harmonisation of operational goals and the elimination of inefficiencies in achieving them. But the animus that underlies alliances, however, is not that of efficiency, but rather that of the political interests that stem from the existence of international anarchy. As such, alliance operations are frequently marked by infighting and competition. NCW might be one tool for alleviating these problems in the hopes of generating a common operating picture or the development of a shared awareness between alliance partners, but the problem is political, not technical.

Information release policies are purposefully inefficient tools in order to protect the information, the sources used to gain it, and the organisations using it from the harm that would result from disclosure to hostile forces: before information can be shared, the 'owner' must be convinced by those desiring the exchange that no harmful effects will take place.[21] Furthermore, because the long-term effect of individual disclosures can be difficult to ascertain, and because the career impact of improper disclosure is so serious, 'commanders often choose stringent release rules to avoid problems'.[22] Thus, information security concerns have dictated separated networks operating at different tempos. As Brigadier-General Gary Salisbury, director of command, control, and communications systems for US European Command, characterised the situation in September 2001:

> How do [combined planners] get these national communication and information needs and fit these into a coalition environment? The bottom line is we are generally operating two different networks at two different security levels. We run our networks at a coalition releaseability level that's basically unclassified.[23]

As Dwight D. Eisenhower remarked, 'Allied Commands depend on mutual confidence.'[24] Like relinquishing command and control, releasing sensitive information is an act of trust between states surpassed only, perhaps, by placing troops under even the limited control of an ally; releasing closely held knowledge places technology, operations, and even personnel at risk.[25] 'Trust involves a willingness to be vulnerable and to assume risk. Trust involves some form of dependency.'[26] The nature of the international environment makes trust exceedingly difficult to achieve, even in alliance contexts. Furthermore, military partners generally exploit dependencies in order to enhance their control over alliance policies. Thus, we can expect that, just as nations have always been unwilling to place their troops under the command of other nations, they will be unwilling to share completely all information they have: 'As close as … Canadian and British allies are in common interests and objectives, there will always be limits to sharing the most highly classified information with these nations.'[27] In the past, this reluctance did not typically jeopardise operations. However, in NCW information is the cornerstone of all action; the existence of separate networks operating at different speeds will have a serious impact on battle rhythms.

NCW, then, will have an enormous bearing on how alliances and particularly coalitions will conduct their operations in the future. The United States is certainly willing to share much of its information with its closest allies, typically the UK, Australia, Canada, and even New Zealand in certain circumstances. However, for the forces of countries not in this privileged club, integration into American networks will be increasingly difficult, depending on how often they operate with US forces and the degree of trust extended to them. Forces not permitted to take part in planning will ultimately be restricted simply to taking orders, and possibly assuming high-casualty or politically distasteful roles.[28] Multinational operations may become more and more circumscribed, and military cooperation, perhaps even with America's most privileged partners, will be accepted only under the most restrictive circumstances. The United States is unlikely to hamstring its own military forces or to slow its implementation of NCW given the perceptible benefits. It may decide simply to forego entirely alliance participation.[29] Information release policy may ultimately decide, not only the shape and nature of coalitions, but also whether they exist at all. Finally, American military primacy will probably place additional barriers in the way of information-sharing between states, particularly between the United States and its allies. Armed as it is with the full panoply of information garnered by its world-wide intelligence services, the US will provide more than the lion's share of information to its partners, and will only seek highly specialised intelligence from them. Furthermore, the environment of US military primacy itself will generate increasing distrust amongst America's partners as the role of independent national interests in shaping policy becomes stronger.

As information becomes more central to modern operations, the shadow of unilateralism will loom heavily. States will continue to share information amongst themselves, but perfect transparency in the form of seamless interoperability will be impossible. Information is simply too central to the competitive advantages offered by NCW to be jeopardised by automatic disclosure. Such disclosure may happen on a case-by-case basis, depending on the nature of the conflict and the partners with which the US is cooperating. But the dictates of sovereignty will ensure that seamless interoperability will remain confined to the realm of the speculative.

4 Naval networks in the coalition environment

Coalition networks received their first operational tests with the commencement of the War on Terror in Afghanistan in 2001. These networks in general worked very well, and helped the US to manage diverse coalition partners with widely varying levels of technical and professional capabilities, as well as political commitment. That said, these were high-end allies. In this respect, the success of operations in Afghanistan masks a broader set of obstacles.

Tactical, operational, and strategic issues confronting networked coalitions

Research on coalitions and networks is particularly intense in professional military education programmes. While there are few common themes amongst these papers, military students, many of them writing on issues they confronted while serving in a variety of coalition operations, are in general agreement that NCW poses a significant threat to coalition operations as the US moves decisively to integrate IT into its operational concepts.[1] The challenges posed by IT to coalitions exist at all levels of warfare. For example, many analyses of NCW in a coalition environment suggest that the problem is largely one of poor systems integration, or the result of a general lack of capital investment in particular types of technology.[2] Other authors conflate the issue of coalition interoperability with that of joint-service interoperability.[3] Underlying tactical approaches to the problem is the assumption that the proper adoption of technology and associated doctrine will be sufficient to address the problem of information exchange within coalitions.[4]

Indeed, by 2003 information technology seemed to be increasingly complicating coalition operations, rather than simplifying them. One analysis of CENTCOM operations in Afghanistan and Iraq that year noted that American planners were dealing with more than 84 different coalition networks. Only 26 of these had secure Boundary Protection Services (BPS), the fundamental basis of information security. Needless to say, interoperability between this wide variety of networks was extremely variable, and mostly non-existent. As such, information exchange between members of the coalition was often a sluggish affair.[5]

Some of these problems can be ascribed to technical difficulties in linking networks together. Others can be traced to differences in procedures for issuing and formatting intelligence for dissemination, hampering the process of knowledge management. The most common complaint amongst coalition partners, however, is in terms of the protocols regulating information release, an issue that affects the tactical, operational, and strategic levels alike. As early as 1996, American intelligence officials identified this as an issue that would complicate or jeopardise military collaboration between the US and its partners. For this reason, the Director of Central Intelligence issued a directive ordering specific changes to the handling of intelligence and its sharing with, among others, foreign governments and agencies. DCID1/7 required that intelligence be formatted for easy distribution to all users, including foreign elements cooperating with the US on common security objectives. It argued that caveats such as NOFORN, WNINTEL, and various REL TO (releasable to) or REL[6] overly complicated intelligence-sharing, especially since these were usually applied with relatively little assessment as to their necessity. DCID1/7 attempted to resolve these problems by eliminating the various caveats and control markings, and suggesting methods by which even highly sensitive intelligence could be produced that would ultimately be releasable in a coalition environment.[7]

Coalition complaints about the continued application of these caveats continued long after this new policy was issued.[8] As one study put it, 'it is highly unlikely that raw real time data from strategic sensors would be made available to coalition partners. Rather only track information would be provided.'[9] (Track information is processed information from radar returns, showing the 'track' that a target on a radar screen is following, and other information associated with its identity.) Nor did there seem to be a particular technical solution to sort out who received what. At that stage, filtering technology had 'not been designed to differentiate between data releasable to one nation from that releasable to another…'.[10] Given the shortage of foreign disclosure officers, and the disparate nature of the coalition cobbled together by the United States to fight the War on Terror, this problem is unlikely to be solved in the near future. Thus, despite a recognition of the problems intelligence dissemination was causing at the operational level, and a declared need to expand cooperation with America's foreign partners, the demands of national security have continued to frustrate information exchange.[11]

Efforts to network coalition partners

With the introduction of computer networking technology and the benefits associated with it, the United States and its principal allies have established a number of new forums for discussing and resolving the pressing problems of information exchange. The Combined Communications and Electronics Board (CCEB) and the Multinational Interoperability Council (MIC) both took on the challenge of improving the principal Western states' capacity to exchange information amongst themselves.[12] In its *Coalition Network Strategy*, released in June 2004 and updated a year later, the CCEB seeks to move its members away from multiple bilateral network connections and towards a single coalition domain 'supporting information exchange at different security classification and releaseability levels between different coalition partners and communities at all levels of command'.[13] This is an ambitious aim given the problems associated not only with user authentication and information assurance, but also alliance politics. Coalition networks suffer from access problems because of the large numbers of individuals from many different organisations and nationalities that must be linked together. Because of their size and composition, coalition networks tend to be more vulnerable to breakdowns in communication links, suffer from poor confidentiality in terms of their data and are troubled by complex configuration management due to the many different types of computer systems and software applications that must be linked together.[14]

Strategic or national domains permit information sharing within a nation's borders and thus tend to be highly secure, rigidly configured networks that permit little or no access for external partners. Allied or bilateral domains permit a certain degree of sharing between national domains, based as they are on pre-established information exchange agreements. Many are permanently established networks that 'tunnel' into each other, permitting the exchange of e-mail and sometimes web browsing. Information security is difficult to build into coalition networks because of the often ad hoc nature in which coalitions are formed and the tendency of nations to move in and out of them. Thus, coalition networks are frequently stand-alone systems shared between the various partners.[15]

Given the advantages of networks in general to military forces, higher levels of information security on a network permit greater degrees of shared awareness and collaboration. While all members of the coalition may have access to the 'track information', those that also possess the intelligence that has cued the sensors, the raw data they are generating, and the details of plans under development can make more refined judgements on the nature of that track and the likely actions that may have to be performed. However, such information on coalition networks has occasionally been absent or of dubious quality.

Of course, securing unity of effort has always been the principal challenge confronting coalition commanders. The standard solution for allies has typically been geographical dispersion between forces to obviate the need for complex coordination or reduce the possibility of friendly-fire incidents. Within a networked environment, information technology should in theory ease the challenge presented by getting the militaries of different nations to conduct combined operations effectively. Examining the annual reports of the MIC, however, one senses the frustration among military officers confronting this challenge. In the first report, in 1999, reference was made to 'low-level personnel' making decisions on intelligence sharing that result in major operational effects on coalition actions.[16] The following year, the report noted continuing difficulties in information exchange, and called for the implementation of a series of checks to determine where the problems were. Divisions were also noted between partners, with Australia arguing that the recommendations under discussion were 'not aggressive enough', while the US and the UK argued for caution given the need to properly control information.[17] The following year's report noted the 'continuing challenge to draft disclosure policies that meet a variety of different national disclosure policies and processes in a multinational sharing environment'.[18]

As an interim step, both the MIC and the CCEB have sought to establish standards to move coalition networks towards freer exchange. The CCEB established a two-tier framework for classifying networks and their associated levels of security. Thus, Tier One networks are those with BPS, enabling connections to national command and control systems. Tier Two networks possess no BPS, and thus require a stand-alone coalition network in order to bring partners to at least some level of shared awareness.[19]

Two network structures have been developed that reflect this bifurcation. The Coalition Enterprise Regional Information Exchange System (CENTRIXS) is an operational level network, supporting regional commanders and their staffs at a variety of security levels. CENTRIXS permits the exchange of a common operating picture, e-mail with attachments, a common intelligence picture, web-enabled services, and secure voice links. Currently, CENTRIXS is a family of wide-area networks that evolved from the Coalition Wide Area Network (COWAN) first used in Rim of the Pacific (RIMPAC) exercises in the late 1990s.[20] By the time *Operation Iraqi Freedom* was launched in 2003, the series of COWANs had become CENTRIXS systems. CENTRIXS Four Eyes (CFE) has replaced COWAN A, networking the US with the UK, Canada, and Australia. CENTRIXS GCTF (Global Coalition Counter-terrorism Task Force) replaced COWAN C, and has nearly 60 members. CENTRIXS XX permits information sharing between Australia, Canada, the UK, and the US within CENTCOM, while CENTRIXS 0 is a US-only domain.[21] CENTRIXS J permits sharing between the US and Japan during RIMPAC exercises, and CENTRIXS R does the same for the US and South Korea.[22]

CENTRIXS is an operational network; the GRIFFIN system, by contrast, is a secret level wide-area network that permits collaborative planning at the strategic level of command between the US, the UK, Canada, and Australia. As a permanently deployed network, GRIFFIN allows for the proper accreditation of users and the standardisation of applications. As such, it permits information sharing up to the secret level between national domains. Given its permanent nature, a high degree of bandwidth can be employed by the network, allowing rapid and timely access and posting of information.[23]

Operational use of networks in coalitions: Australia and Canada in the Gulf

These networking technologies were used operationally for the first time in 2001, with the formation of the coalition to fight the War on Terror. Both Australia and Canada have participated extensively in this ongoing campaign. Although each has adopted different roles in the War on Terror, and each took very different paths in terms of operations in Iraq (Canada abstained, while Australia has committed forces), each country has fielded relatively similar capabilities in operations in the Middle East. In the naval sphere, both provided task groups composed of frigates for sea control operations. Australia supplemented its frigate deployments with the periodic deployment of amphibious ships, the landing platform docks (LPDs) *Manoora* and *Kanimbla*, and Canada supplemented its frigates with resupply vessels and the destroyers *Iroquois* and *Athabaskan*. Each navy has a long tradition of interoperability with the US Navy dating back to the Second World War. Both navies have operated alongside the US in the Persian Gulf since the early 1990s.[24]

Canadian naval operations fell under Operation Apollo; Australian operations for Operation Enduring Freedom were code named Slipper, and those supporting Operation Iraqi Freedom were named Falconer. Each navy conducted similar missions in separate regions, although Canadian ships occasionally supported Australian operations in the northern Persian Gulf. Ultimately, the Canadians took control of the 'Leadership Interdiction Operation' (LIO) in the Southern Persian Gulf, the Strait of Hormuz, and the Gulf of Oman. The Australian navy operated in the Northern Persian Gulf, where it had patrolled in three separate deployments since 1996.[25] There, it continued to conduct maritime interdiction operations (MIO) and general sea control tasks. Despite the similarity between their missions, each navy's operational area was significantly different. The northern Persian Gulf is a shallow body of water, hemmed in on three sides by the Al Faw peninsula, the Arabian Peninsula, and Iran. Besides the local knowledge of the area built up over ten years of operations there, the shallow drafts of Australia's LPDs and its *Anzac*-class frigates made the Australian Navy an ideal force to operate in the area. The Canadian Navy worked in a much larger area, first in the Arabian Sea, and later in the Gulf of Oman and southern Persian Gulf. This is one of the busiest shipping lanes in the world, moving 30 per cent of the world's annual oil shipments. More than 450 vessels transit the area daily. These ranged from small wooden *dhows* to supertankers, typically generating nearly 6,000 radar contacts on a regular day.[26] Given their different operational environments, the Australian Navy conducted a traditional close blockade of the Iraqi coastline, whereas the Canadian Navy's operations were oriented towards sea control and distant blockade.[27]

The role played by SIPRNET

Ensuring technical interoperability between naval task groups during the Cold War often involved ensuring that the proper cryptographic keys and the right frequencies were coordinated, so that secure radios could communicate with each other. The emerging digital environment has complicated this process considerably because it requires the installation of hardware and software (including the proper version and latest updates), firewalls, accreditation, IP addresses, connectivity paths and processes, and sufficient communications bandwidth to carry the burgeoning traffic exchanged between forces.[28] Furthermore, ensuring that all of this is present has expanded beyond the technical and procedural realm of tactical interoperability and into the realm of strategic policy governing relations between states.

The prime example of the strategic impact networks play has been the growing importance of the US military's SIPRNET for managing information and running global operations. In 2003 former US Fifth Fleet commander Admiral Thomas Zelibor elaborated on his experience with using the SIPRNET in his carrier battle group during the Iraq war, describing it as the evolution of a 'knowledge web' that contained the operational 'ground truth'.[29] COWAN performed analogous functions for the coalition, but one Canadian ship's captain, reflecting on Zelibor's observations, noted that COWAN was 'not where the real battle is being fought, at least not yet, and perhaps never', as it only 'offered a small and sometimes opaque window into the total situational awareness of the USN's battlespace'.[30] Indeed, despite the connections between coalition-wide area networks and the SIPRNET, many coalition officers continue to express some frustration over the difficulties created by the use of separate national and coalition networks because of the demands of national security. One Australian liaison officer working within CENTCOM described the 'abject failure' encountered in trying to cross-register US SIPRNET user accounts as CENTRIXS X (Australia/UK/US) accounts.[31] Both Australian and Canadian officers remarked on the need for US command oversight, often from the highest levels, so that network interoperability can be made more effective with American ships.[32] The transfer of essential planning information to coalition partners occasionally fell through the electronic cracks between networks as units sought to establish who was responsible for releasing information, or because units, challenged by the pressure of operations, failed to post information quickly enough. In this regard, US military forces naturally operated at higher levels of efficiency because they could look up the information on SIPRNET.[33]

Australia was able to negotiate the installation of a SIPRNET terminal on its LPD *HMAS Manoora* during the autumn of 2002. The terminal was placed in a compartment aboard the ship crewed exclusively by US personnel. Australian Rear-Admiral James Goldrick noted that he 'could not have operated as [the Maritime Interdiction Force] commander without it, so reliant have C2 processes become on SIPRNET e-mail and chat, particularly the latter'.[34] Despite the limited duration and access of the Australians to SIPRNET, the fact that it existed at all weighed heavily in the minds of Canadian officers lacking similar access, concerned that the Canadian decision not to participate in Iraq had somehow moved them to the outer circle of allies. As one put it:

> the true test of whether or not you were an inner circle member: are you on SIPRNET? ... The only coalition partners that have access to SIPRNET now is [*sic*] the Brits and the Australians. ... That to me is the dividing line between those on the inner circle with the US. Because the US does all its [operational planning] on the SIPRNET.[35]

'Concentricity' of access

'Circles of access' were reflected in more ways than simply 'network permissions'. Most officers interviewed perceived the US-led coalition as structured in a series of concentric circles of access, with the US at the centre position. The UK occupied the circle closest to the US, followed by other 'anglo-sphere' nations, other NATO states, and then the rest of the coalition.[36] CENTCOM reinforced this structure by insisting on dealing with each coalition member bilaterally, rather than seeing the coalition as a coherent entity. The perception of coalition naval officers serving in the Gulf was that the US did not want to get into a 'NATO-type situation' where everything from strategic policy to operational planning and tactical targeting had to be negotiated in advance. Former commander of the Canadian Joint Task Force South West Asia Brigadier General Angus Watt noted: 'If you are a coalition member, you plug into the US agenda and if you don't want to follow [it], you ain't a member of the coalition. It's that simple.'[37]

The US was clearly sensitive to any perception that states were not being treated appropriately, and worked hard to ensure that it dealt with each nation in a similar way irrespective of its contribution to the war effort.[38] Nevertheless, the concentric circles of access became increasingly apparent after the Canadian government delayed committing forces to operations in Iraq in late 2002 and early 2003. This contrasted with Australia's willingness to discuss options very early in 2002 (Australia was included in an Operation Iraqi Freedom planning cell in October that year).[39] The Canadian military was ultimately able to convince the government in Ottawa to establish a liaison team to discuss possible Canadian participation at the end of November 2002. Following this, 'the Americans appeared to open the doors very wide and gave [Canada] a lot of information about their intentions…'. However, as Canada continued to delay its decision, 'the doors weren't closed, but you could feel them closing'. One of the ways this became apparent was in the nature of the information that was provided to Canadian liaison staff. Information within US headquarters is circulated in the form of PowerPoint briefing slides. These briefs are often extremely large, sometimes numbering 1,000 slides or more as each decision point, 'branch', and 'sequel' operation has its own set of hidden and embedded slides. The detail that Canadian officers were allowed to see was progressively restricted until it reached the standard coalition releaseability level, sometimes referred to dismissively as the *Reader's Digest* version.[40]

The physical layout of CENTCOM, both in Florida and Qatar, also reflected this segmentation of information. Outside the Florida headquarters in Tampa is a 'trailer park' of coalition members. This was also reflected in the CENTCOM Forward HQ in Qatar which maintained a 'Friendly Forces Co-ordination Center' outside the main building in an large inflatable tent. While some coalition members, principally the UK, Australia, and Canada, operated as liaison officers and embedded planners within the HQ, all others were restricted to the tent. One Australian liaison officer working at the Qatar HQ claimed, 'physically and in a cognitive sense, I was separated', from the other Australians working in the Co-ordination Center.[41]

Such arrangements within headquarters were not new. Indeed, information within a military headquarters is often controlled even between planners from the same country and service. The physical barriers are replicated electronically, in that it is easy to provide information to the SIPRNET, but much more difficult for coalition partners to get information back out of it. As one Australian liaison officer conducting planning within NAVCENT HQ for Operation Iraqi Freedom noted, 'any of the work I would do, would be done on [a] stand alone [system], and then loaded up to the SIPRNET'. Information was downloaded for his work by US officers, and only then passed to him: 'NAVCENT HQ maintained this structure through to execution'.[42] The difficulty coalition members face is that, unless American users cue them to request specific products, the material provided is likely to be of little value, and come too late. In a large coalition, where partners are all requesting information from SIPRNET, the sheer number of requests quickly exceeds the available resources to process them. Those closest to the centre will be best served.

SATCOMs and information access

In the Gulf, inadequate satellite communications created a second crucial bottleneck for NCW. As many as six separate networks could be running on a single ship, including classified and unclassified national networks, a coalition network such as COWAN or CENTRIXS, the GCCS providing operational level situational awareness, and tactical data links like Link 11 and Link 16. Access to these networks could only be assured through satellite communications (SATCOM) channels. Many coalition members failed to provide these channels to their forces, or used 'dial-up' access to commercial communication satellites through INMARSAT, rather than paying for continuously running, leased channels.[43]

In this field, Canada had advantages enjoyed by no other coalition member. Thanks to the towed-array sonar technology it developed during the 1980s, the Canadian Navy has long been involved in the over-the-horizon networking efforts of its US counterpart.[44] In addition, Canada helped fund the US Navy's Fleet Satellite Communication (FLTSATCOM) satellites, and has eight national channels there. By 2001, Canada had also leased 12 continuously running INMARSAT channels, six of which were given to the Navy for use in Operation Apollo. Each of these channels was further multiplexed,[45] allowing separate networks to run on each channel.[46] This gave Canadian ships a communications capability that rivalled that of some larger American vessels.

At the time, Australian capability was much more restricted. Only the larger LPDs maintained a continuous connection to INMARSAT. Moreover, only the LPD could multiplex its SATCOM connections. The frigates were limited to dial-up access at particular times of the day, or during the execution of operations. The bandwidth of these connections was also significantly smaller than that of Canadian ships, at 64 Kbps as opposed to 128 Kbps.[47] While the Australian Navy briefly enjoyed the services of SIPRNET, it had to sacrifice one of its two channels on the LPD to gain it as American information security protocols demanded a dedicated channel to carry SIPRNET traffic.[48]

Bandwidth scarcity also affected US ships. While primary units such as the carrier and some cruisers enjoyed larger amounts of bandwidth, the increase in the number of American ships in the Gulf as the war with Iraq drew closer meant increasing competition for a fixed resource.[49] From a coalition perspective, bandwidth scarcity is a serious issue affecting most navies. For example, in the major biannual coalition exercise in Asia, RIMPAC 2004, the US Navy managed five separate coalition network domains as well as many national ones.[50] Bandwidth limitations inevitably mean that certain networks, especially coalition ones, will be monitored by American crews less frequently, and given lower priority than the SIPRNET.

Coalition information sharing

Despite the challenges, Canadian and Australian officers described the sharing of information between them and the US as generally satisfactory, and all claimed that they had enough information to conduct their operations. The nature of the threat in some instances dictated the amount of intelligence that was shared (and sometimes the lack of it). Intelligence on the movement of small boats and aircraft, for example, was lacking early in the campaign. Canadian Commodore Drew Robertson, who commanded the defensive screen around American amphibious ships in 2001, was extremely impressed with US willingness to share operational planning details with the Canadian task group: 'we knew what they were doing ... I knew every day when they planned to go to and from the beach, or to and from the various operating areas so that I could organise our ships appropriately.'[51]

Still, the separation of networks also presented problems, especially in terms of operational planning. Robertson's task force lacked both e-mail and voice connectivity with the first amphibious group it escorted as the Americans did not have any COWAN terminals. This meant that US plans could not be shared with the Canadian task force until they had been finalised, and Canadian planning could not begin until the plans had been sent. In the Arabian Sea in 2001 this was a minor annoyance, but Robertson noted that 'in certain situations there won't be time for that kind of wheel spinning ... Those kinds of inefficiencies can lead to real problems.'[52] Later in the campaign, Canadian Commodore Eric Lerhe led an effort to populate COWAN web pages with information to increase the speed and efficiency of coalition force planning at sea. However, time pressures often meant that the US was unable to maintain its COWAN pages effectively as well as its SIPRNET pages. Lerhe concluded that 'the bottom line is we are going to have to, for every operation, pull the levers, kick the tyres, and scream to get everybody working. But people who hang their hopes on [multilevel security], I just don't think they are operating realistically.'[53]

Classification barriers to information release for coalition networks were particularly evident in the interaction between Special Forces and the coalition. Both the Australians and Canadians experienced at least one incident each that nearly resulted in fratricide on US Navy SEAL teams. In each case, SEAL teams had failed to keep even their own national forces informed of their activities, leading to widespread confusion as to whether the SEAL team involved was a friendly force or an enemy target. Coordination over radio nets between coalition forces ultimately resolved the problem, although in the Australian case not before authority to open fire had already been given.[54] The significance of these events points to the challenge that highly classified networks present to coalition operations should information barriers for more conventional operations come to resemble those currently present on special forces networks.

In general, intelligence sharing is conducted on special-access networks between the American, British, Australian, and Canadian coalition partners. However, during operations in the Gulf between 2002 and 2003 the quality of some of this material was often uncertain. In terms of the LIO being conducted by the coalition against Al-Qaeda, good intelligence on intended routes was often lacking.[55] So-called 'actionable' intelligence was also suspect. As Robertson noted:

> I found it useful that actionable intelligence could lead to nothing because it reminded me and it reminded my COs that actionable intelligence isn't a certainty, it's a probability … You wouldn't want to over-react and find that you had just done harm to some poor merchant mariners of the region.[56]

Even basic track information shared on these networks was often not accepted as accurate. For example, the GCCS is used to manage track information globally. In Lerhe's task force, reservist specialists double-checked GCCS data with information found on port web sites. Significant discrepancies between the GCCS data and these open sources led to considerable scepticism of the GCCS system within Lerhe's task force from time to time.[57]

Coalition forces developed a series of databases to track the enormous amounts of shipping passing through the region. These databases were important for three reasons. First, the maritime task forces in the Persian Gulf, the Gulf of Oman, and the sea around the Horn of Africa did not possess adequate resources to interpret properly the data they were collecting. Each task force thus shared its database with the others and with NAVCENT HQ, where naval intelligence specialists could 'mine' it for more specific data. Second, the sheer number of vessels passing through the region meant that the coalition had to be very selective in terms of which vessels were boarded. Boarding ships that had already been cleared was an obvious waste of resources. Third, these databases helped the coalition to establish the appearance of a professional and competent boarding regime. Convincing ships' masters that boardings did not take place on a whim increased trust in the coalition and led to a higher compliance rate. This was true even for the small boats transporting economic migrants between Pakistan and the Gulf States. When boat captains realised that coalition forces were only interested in determining the presence or absence of terrorist suspects amongst their passengers, they became much more cooperative.[58]

The sharing regimes that were ultimately established between coalition partners resulted in considerable synchronisation, innovation, and self-adaptation, as the theorists of NCW had anticipated. The importance of this self-adaptation became apparent during the naval gunfire missions HMAS *Anzac* conducted with Royal Marine units on the Al Faw peninsula on 21 March 2003. The Royal

Navy ship HMS *Chatham* also participated in this mission, but a misfire interrupted its support for the Marines. *Anzac* was able to take over the fire mission it had been following over the network, having already entered all the fire control data into its own system.[59] The increasingly vital role played by networks in sharing information amongst partners was summed up by Lerhe, who noted:

> What's your level of tolerance for going into the ops room and saying 'Tell the Italian ship *Euro* to go north and intercept that ship,' … and my guy turning around and telling me 'Sir, we don't have comms with the *Euro*.' … Sure it's extra work, but COWAN is 100%. What's the alternative? Is 96% really the alternative? No! Because Al Qaeda's going to be on the 97th. So it's all or nothing.[60]

The human in the loop: liaison

Given the organisational and electronic challenges of sharing information amongst coalition partners, the human element was often decisive in making the growing electronic environment effective. Liaison officers, long employed to coordinate coalition operations, played a critical role in ensuring that information gaps did not persist. So vital was this liaison function that the most important members of a task force's team were often sent to ensure proper communications between coalition units.[61] The location of liaison officers within a foreign headquarters was a key consideration for national commanders when deploying them. A liaison officer's value could also be enhanced by the roles they played there. Taking on planning duties within a foreign headquarters reduced the 'burden' of liaison officers on US forces.[62] Recalling information age sociologist Manuel Castell's observation about the varying utility of network nodes, a liaison officer's 'value' was enhanced when they became useful to the Americans as an embedded staff member '… And so all of [the Australian liaison officers] picked up, to a certain extent, tasks within the headquarters that they were attached to.'[63] While status as an embedded planner also increased the quality of information these officers sent back to their own forces, that information was very much dependent on which aspect of the plan they were allowed to work on. Liaison officers located on the 'fringe' of activities might get a good picture of that fringe, but little else. Moreover, the US command could easily sideline liaison officers if they failed to perform their planning responsibilities adequately, or if their political utility to the US declined. Still, within the large US headquarters, with personnel continuously moving in and out, coalition liaison officers could contribute substantially to the 'corporate knowledge' necessary for an effective operational plan.[64]

Rules of engagement: intersection of strategic and operational policies

While coalitions present *operational* problems for the US, they present *strategic* problems for its partners. These states' politicians may fear that they are too closely aligned with American policy, and the need to give America operational control of their forces challenges their sovereignty.[65] As American planning for operations against Iraq intensified during 2002–2003, the concern of states opposed to such action, but supportive of American policy in Afghanistan and against terrorism in general, posed both strategic and operational problems.

For coalition partners, the question was how they could continue to support the US in its War on Terror while opposing American plans for Iraq. For the US, the question was how to structure the two operations so that neither would suffer because of differences in strategic policy. On the land and in the air, the issue was fairly straightforward as the theatres of operations were widely separated and there was no possibility of confusion between them. The situation was different at sea. Since late 2001, coalition units had been operating within the American's Task Force 50 in support of Operation Enduring Freedom in the southern Persian Gulf and the Gulf of Oman, the same area in which Operation Iraqi Freedom naval operations would also be taking place. TF 50 included both coalition forces conducting counter-terrorist leadership interdiction operations and general sea control and escort duties, and US Navy Carrier Strike Groups that would conduct air operations over Iraq.[66] The solution was the creation of two separate coalition task forces: one CTF 150 – supporting Afghan operations and commanded by Europeans – operating off the Horn of Africa, and the other CTF 151, conducting counter-terrorist leadership interdiction operations in the Gulf of Oman, under Canadian command. This permitted a 'clear separation of activities between the overt warfighting of Operation Iraqi Freedom and the picture compilation and maritime interdiction of Operation Enduring Freedom'.[67]

The creation of CTF 150 and 151 highlights how strategic policy differences within coalitions (here in terms of differing national policies towards the danger Iraq presented to international stability) affected the management of military operations. These manifested themselves in terms of distinctive ROEs, which were ultimately managed by coalition commanders using the networks they had already established. Indeed, ROEs became as critical an issue in the shaping of operations as connectivity and capability.[68] In anticipation of action, coalition commanders created hypothetical scenarios that permitted all parties to explore what they could and could not do in light of their national ROEs, thus enabling the early assignment of tasks and the positioning of forces. This allowed differing ROEs to be 'blended' together, enabling coalition forces to achieve their maximum potential without violating any partner's strategic policy.[69]

While ROEs are critical in all operations, in littoral environments they can pose delicate challenges. The Persian Gulf and the Gulf of Oman are highly complex in terms of their environmental features and political geography, as well as the maritime traffic passing through the region. Maritime borders are disputed, radio communications are difficult at the best of times, linguistic and cultural challenges confront extra-regional forces operating there, and the relatively confined nature of regional waters amplifies the threats posed by submarines, anti-ship missiles, and the many small craft operating in the area. According to Lerhe, divergences in ROEs

> meant some nations would not react as robustly as US forces ... many contributing nations lacked the ROE that would have allowed them to forcibly board ships or capture terrorist leaders and would only assist in such secondary but important tasks as providing surveillance.[70]

Nevertheless, Canadian commanders still utilised ships with extremely restrictive ROEs. For example, Japanese ships escorting their re-supply vessels provided radar data on distant traffic, giving the task force an additional day's warning of approaching targets of interest.[71] The basis for some ROEs was dictated by strategic policies that were unrelated to the Gulf or the conflicts there. Canada's recognition of Iranian territorial waters, limiting the area in which it could operate in hot pursuits, was related to its use of the straight baseline rule for claiming sovereignty over Arctic waters, which is also used by the Iranians for their maritime boundary claims.[72]

Despite the complications of differing national strategic policies, networks enabled naval commanders to stay in close operational touch with each other, and also provided opportunities to discuss sensitive issues privately before they became serious operational problems. Private 'chat boxes' were established to 'express private reservations or concerns ... *candidly*, while maintaining more public chat circuits that were more disciplined and with many participants for rapid exchange of information.'[73] An excellent example of the networked management of ROE was the case of the Iraqi tug *Proton*. The *Proton* was found at anchor in the southern Persian Gulf on 23 March, two days after the start of Operation Iraqi Freedom and the day after Australian forces had discovered a similar vessel loaded with mines in the Khawr Abd Allah. Because mines posed a 'maritime safety' issue, Canadian Commodore Roger Girouard, the commander of CTF 151, felt that he had sufficient authority under his ROE to board the vessel in order to inspect it for the presence of these weapons. No mines were found, though gas masks, atropine injectors, and Molotov cocktails were present, and the crew appeared to behave suspiciously. However, none of these factors made the tug a matter for the LIO. Girouard informed NAVCENT HQ of the discoveries made but was told to release the vessel. He found this request 'strange', but complied. Later, NAVCENT requested that the *Proton* be reboarded; Girouard refused on the grounds that to do so would have contradicted his ROE. Two days later, the *Proton* was spotted alongside a barge also suspected of carrying mines. At this point, Canadian ROE permitted a reboarding, again because of the maritime safety issues associated with mining international waters.

As complex as this episode was, the fact that Girouard was a Canadian commanding a Canadian boarding party lent it a degree of simplicity. Had he been distant from the scene and reliant on a boarding party from another country, the situation might have been even more complex. The incident highlighted not only the value of a network permitting all participants to exchange information, and to communicate securely, but also the significance of the transition from COWAN to CENTRIXS in the region. Canadian commanders initially resisted the transition as it effectively meant that 49 additional nations would be added to the network, with a resultant decline in the quality of information residing there. However, CENTRIXS also expanded the means by which all coalition ships could communicate securely and reliably over digital links. While less robust in terms of access to classified information, in terms of managing the coalition CENTRIXS was superior to COWAN in linking all the players together.[74]

Networking the coalition: social and digital factors

For coalition naval operations in the Gulf, networks were an important enabler in a very traditional naval mission not unlike the gunboat diplomacy familiar to nineteenth-century naval commanders. While the naval operations in the Gulf in 2002–2003 succeeded in that all the missions undertaken were accomplished and no casualties were sustained, the absence of serious opposition raises questions as to how they might have proceeded in a scenario 'more closely envisioned by the proponents of NCW'.[75] As much as networks were critical to the sharing of situational awareness and in mission planning, operations Apollo and Slipper/Falconer were very different from those envisaged by Gartska and other enthusiasts of NCW. The need for information security ensured that there was no 'seamless architecture'. Indeed, the information release protocols of every networked participant engineered just the opposite: a proliferation of networks and thus a proliferation in the number of 'seams'. Virtual electronic borders mimic real national boundaries. Moreover, computer networks have not obviated the need for personal interaction. Indeed, the evidence suggests that, in order for computer networks to function as efficiently as possible, social networks need to be established first. While building an electronic network is a relatively simple matter of capital investment and proper training, creating a social network is a much more complicated task.

Building both strategic and professional trust is a timeless challenge. The fear of abandonment that stimulates cooperation at the strategic level works at the operational level as well. National perspectives directly influence operational thinking. Put simply, US commanders need to win; non-US commanders in the coalition want to make a meaningful national contribution, but they also want to minimise their casualties. Under these circumstances, can the US trust an ally or coalition partner to do what is necessary to accomplish a mission, or are these partners simply operational burdens, there merely to show their national flags? The coalition partner's concern is whether it will be allowed to play a meaningful role, and whether the missions planned by the US will be politically acceptable. The need to accommodate a coalition partner's desire for a significant mission (and thus its influence over an operation) has to be carefully balanced against its capacity and willingness to see that mission accomplished effectively. This is essentially a question of trust. As one Australian commander put it: 'To the USN a new … [foreign] command team was a completely unknown quantity. Only through your actions could confidence be built up with you and your team.'[76] Furthermore, even if trust were established between individual commanders, this accomplishment still had to be communicated effectively upwards to higher headquarters, and downwards within planning staffs. In the Gulf, memos clearly articulating what commanders could and could not do were widely distributed amongst command and planning staff in order to minimise 'second-guessing' during operations. Commanders also met on a regular basis so that their staffs could see how well the leadership got along together. Repeatedly, commanders referred to the great traditions of naval operations, frequently invoking the pre-electronic example of Nelson's 'band of brothers' who fought at Trafalgar.[77]

The Persian Gulf's various coalition networks were undoubtedly successful: they created operational and tactical situational awareness, which was shared effectively amongst partners. However, several caveats apply. High-end coalition partners like Australia, Canada, and the UK have sufficient access to, and professional trust within, the US Navy to guarantee their connectivity, and ensure that other coalition members enjoy relatively similar benefits. In the Gulf of Oman, American trust permitted considerable innovation by the Canadian Navy in developing the coalition network, ensuring continued coalition support for the War on Terror even as the coalition was put under strain by the invasion of Iraq. In this instance, a close ally like Canada was available and capable of leading the segmented operation in the Gulf of Oman – a situation which is not necessarily guaranteed in the future. Again, it is doubtful that other nations could have played similar roles as effectively. Had access and professional trust been absent, cooperation would have been crippled at the outset.

5 The neighbourhood watch

Organisational and political boundaries in NORAD

Air Operations Centers (AOC)[1] are the quintessential network centric organisation; they are, in one account, the 'the controlling hand that is on every weapon'[2] in the air domain. Since the success of the Gulf War in 1991, their use has spread significantly in the running of air operations, so much so, that the USAF now considers them to be a weapon system in their own right.

Few nations have a deployable AOC capability, although NATO has a number of fixed centres. As a result, AOCs are an overwhelming American affair to which nations contribute personnel, typically in the form of liaison officers.[3] In this regard, *Combined* Air Operations Centers, or CAOCs, are operational level affairs into which nation's post military personnel in a plug and play fashion. In order to play, nation's must first be allowed in, and then, either check their strategic reservations at the door, or spell them out in clear and unambiguous terms.

One example where this is not true is in the complex of command and control facilities that has developed over the past 50 years between Canada and the United States, exemplified in the famed North American Aerospace Defense Command, or NORAD, headquartered in Colorado Springs. Strategic concerns on the part of both Canada and the United States have never been far from the conduct of operations there, and have often caused concern on the continued survivability of the organisation. Indeed, one way of looking at the history of NORAD is a continuous effort on the part of the airmen working there (now a joint command) to find work-arounds to the strategic concerns raised by stresses and strains in the political relationship between the two countries.

Networking technology is nothing new to NORAD. From the very beginning, consideration of the North American geographic area as a single battlespace motivated the operational cooperation between Canada and the United States. Air operations in particular, lent themselves well to notions of networked information sharing. Indeed, the first air operation centres were established by Great Britain in the form of Fighter Command in 1936. NORAD has long used aspects of network centric warfare to coordinate responses to threats to North American security. It is this coordination which has proved most controversial.

From the Canadian perspective, ongoing unease over the level of control the United States had on decisions to go to war made NORAD a frequent target of Canadian nationalists who were convinced of American plans to control the country. NORAD was never as remotely controversial for the average American, who rarely concerned himself with the bi-national management of North American security. However, the American military did worry that political interference would undermine the necessary operational coordination between Canada and the US for continental defence. As such, the US developed a series of back doors to ensure that it ultimately could conduct unilateral operations in the event of Canadian objections. The history reflects, in fact, declining levels of cooperation between the militaries, even as the technology that permits such action develops increasing levels of sophistication.

NORAD demonstrates effectively the limits of information sharing between militaries in a fully networked environment. Information sharing in NORAD has been constrained not by access to technology, where the United States has often gone out of its way to assist its Canadian partner. Rather it has been the political concerns of both partners that have posed the most difficult challenges. While the geographic nature of the North American continent has driven the two countries together, a feature reinforced since 9/11 with the rise in importance of Homeland Defence and the subsequent evolution of Northern Command, issues of sovereignty and different perceptions on national security have had clear impacts on information sharing between two organisations that enjoy extremely good professional relationships. Taken in terms of the realm of the possible within the more prosaic CAOC, NORAD has lessons that suggest that the seamless network sought after in terms of pure theory remains chimerical in the extreme.

Bringing order from chaos...

The Army Field Manual 100–20 of 1943 notes:

> The inherent flexibility of air power is its greatest asset. This flexibility makes it possible to employ the whole weight of the available air power against selected areas in turn; such concentrated use of the air striking force is a battle winning factor of the first importance. Control of available air power must be centralized and command must be exercised through the air force commander if this inherent flexibility and ability to deliver a decisive blow are to be fully exploited.[4]

In many ways, the wisdom of this approach to the use of air power, employed both by the RAF and the USAF during the Second World War, including during their Combined Bomber Offensive, took nearly another 50 years before it was effectively reapplied in modern combat. The history of air power following the Second World War is that of persistent bureaucratic struggles over the application of air power between service champions, and sometimes struggles even within a service. However, by 1991, air operation centres exceeding the capabilities to coordinate large scale air campaigns of their Second World War counterparts had begun to appear in operations. The growing importance of joint control of all aspects of modern campaigns provided the stimulus for this rediscovery, as did the role of Goldwater-Nichols act in forcing the services to overcome their inertial bureaucratic tendencies.[5]

In essence, CAOCs permit the centralised planning, direction, and control of the large number of independent organisations and their constituent parts to accomplish air tasks in support of the overall campaign plan guiding a military operation.[6] 'The CAOC provides the commander the capability to direct and supervise the activity of assigned, supporting, or attached forces and monitor the actions of both enemy and friendly forces.'[7]

In order to accomplish this mission, CAOCs are functionally divided into a number of different internal organisations including Strategy, Combat Plans, Combat Operations, Intelligence Surveillance & Reconnaisance, and Air Mobility. The aviation specialist David Fulghum notes that these functional areas can be roughly categorised as 'finders', 'deciders', 'shooters', and 'supporters'. Basically then, the CAOC seeks targets, decides their importance and any actions that might necessarily follow, and matches targets to assets. Underneath these glamorous roles, a series of more mundane but no less important tasks that go to ensuring the ongoing operational viability must also be performed such as plans, administration, and logistics. Still further, 'special technical operations' will also be undertaken by CAOCs, such as network attack, information warfare, and other technical specialities like electronic warfare. These highly classified domains are typically hived off from the main CAOC in specialised cells that may even be in entirely separate facilities, though still under the direction of the centre itself. As at sea, chat is the principal means of coordinating action in this diverse group of actors.[8]

The centre of this hive of activity is the so-called 'Battle Cab' where the many information flows are centralised for assessment, planning, and control. This is typically also a highly classified environment where information is processed at a variety of classification levels.[9] The archetypal CAOC is located on the Al Udeid Air base in Qatar, which runs air operations for both Operation Iraqi Freedom (OIF) in Iraq, and Operation Enduring Freedom (OEF) in Afghanistan. It is easy to overlook the fact that this one centre manages air combat operations involving fighters, bombers, and their supporting air craft, together with air lift missions supporting both OIF and OEF over a geographical area spanning 5,600 km, from Pakistan to Africa.[10]

This enormously complex task that crosses many different combat specialisations ultimately resulted in the establishment of the AOC as a weapon system, in and of itself, by the USAF early in this decade. The value of the AOC is its ability to bring order from what would in any other circumstance be a highly chaotic undertaking. However, the very complexity of the task meant that in the absence of centralised direction, each AOC would develop unique command and control methods and approaches to running operations. This lack of conformity was as much a threat to operations as the absence of any centralised direction in the first part. As AOCs in peacetime would be less than fully staffed, during a crisis, the lack of conformity and standardised operating procedures would mean that incoming staff would lack the understanding of how the organisation functioned, hindering the mission. In turn, this would generate an enormous training burden on the organisation, just at the point when full capacity was required. As such, the formalisation of procedures and technology by establishing the AOC as a weapon system would permit airmen to be trained as part of their ongoing professional development, ensuring the availability of competent personnel in times of crisis.[11]

At present, the USAF operates five static 'Falconer' AOCs at the already mentioned Al Udeid Air Base, Osan Air Base in Korea, Davis Monathan Air Force Base to Arizona, Ramstein Air Force Base in Germany, and Hickam Air Force Base in Hawaii. In addition to these, there are an additional five 'functional' Falconer AOCs that perform very specific duties: Tyndall AFB in Florida, Elmendorf AFB Alaska and Cheyenne Mountain AFB Colorado all support homeland defence through their NORAD duties, Vandenberg AFB California performs space-related functions, and Scott AFB Illinois handles mobility, air lift, and tanking functions. In addition to these American AOCs, NATO maintains ten functionally similar CAOCs.[12]

The engine that supports the complex operations of American AOCs is the Theatre Battle Management Core System (TBMCS), a conglomeration of nearly 80 unique command and control systems formed through the merger of three older systems: the Contingency Theatre Air Planning System, the Combat Intelligence System, and the Wing Command and Control System. TBMCS is an advance over these older systems in its ability to generate Air Tasking Orders (ATOs), the backbone of all modern air operations, with three times the level of information on sorties and targets, in one-half the time, using one-third the planners.[13]

TBMCS is a true representative of the merger of force and information in the air domain. Considering Table 5.1, TBMCS permits air planners and controllers to merge information from a large number of domains, permitting a high level of control and fidelity in terms of the desired effects.

As can be seen, the production of the ATO requires the fusion of a large number of information feeds, together with the cooperation of an equally large number of organisations. If this process is managed too centrally, the resultant ATO may not make full use of the capability any given organisation can provide. As some have pointed out, the goal of ATO production can become the object, rather than achieving the operational goals that have been established. The organisational complexity of ATO production has in the past produced

Table 5.1 Theatre battle management core systems capabilities

Air campaign planning	Execution management: replanning
Air space deconfliction	Close air support tool
Theatre air planning	Time sensitive targeting
Joint defence planning	Situation awareness
Weather	Intelligence: target and weaponeering module
ATO	

Source: www.mctsss.usmc.mil.

mechanical procedures that stifle flexibility and agility in planning and execution.[14] TBMCS permitted the USAF to take what had been highly stovepiped production and assessment tools and unify them into a single coherent package in which information flowed freely between organisations.

Further, the same tool which has advanced air operations planning has also permitted greater levels of jointness through the development of a real time common operational picture that can be shared over the GCCS system. This has, in turn, enabled the 'coordination of precision fires, safe passage zones, and near real time air raid alerts'.[15] The complexity of air operations planning and execution is further enhanced through the integration of the Automated Deep Operations Coordination System (ADOCS), a joint mission management software tool used by all three services in order to plan joint fires in the battlespace.[16]

The focus for future development in the digital management of information within the AOC is to further decrease the remaining organisational barriers that still exist. Despite the clear advances that AOCs made between 1991 and 2003, senior officers within the USAF still note the problematic effects that organisational hierarchies and divisions create in the management of information for the conduct of operations. LGen. Ronald Keys, Deputy Chief of Staff for Air and Space Operations called OIF a 'war of neighbourhoods' within the CAOC.[17] General G. Martin, Commander of NATO's AFNORTH called for the need to take a 'fragmented pie' and turn it into a seamless continuum.[18] Keys noted further, networks

> should now be interconnected, and integrated in order to form 'city wide nets,' … we need a global commercial Internet type of capability. No matter where I am or in what platform, I should be able to log on to this net.[19]

The hope here is that an Internet centric approach to data, together with the standardisation of procedures within CAOCs will have similar effects on decision making within them as the Internet in general has enabled the social interaction of individuals. In sum, these will create a more flexible and rapid approach to the planning and execution of operations.[20]

These expectations are beginning to manifest themselves in some ways already. The ATO production process has been revolutionised since 1991 into something far more dynamic. What had been a highly bureaucratised process, requiring days of advance planning, in 2003 was able to accommodate on the spot changes to routing and targeting. Based on planner's understanding of what specific aircraft were carrying in the course of their missions, aircraft were able to be re-tasked to accomplish other missions. For example, F-16CJs conducting SEAD missions with AGM-88 HARM missiles, also carried JDAMS and CBUs. Mid-mission, these aircraft were able to be diverted to undertake other air to ground tasks.[21]

The growing use of 'kill boxes' has also been enabled by this new approach to data. After criticism by the US Army on the lack of responsiveness of the USAF for CAS missions, both the Army and Air Force revisited how liaison was handled and how forward air controllers were trained, in order to ensure that the procedures and terminology used by them were identical. During OIF, Iraq was divided up into a series of 30 mile by 30 mile boxes. Depending on where the box was located in a specific period of time, different rules of engagement could be applied to it. If troops were located nearby, different ROEs were employed. If the kill box was in front of the forward control line (FCL), then 'eyes on target' were necessary to engage.[22]

In some respects, this operational 'addressing' of Iraq's physical geography works in the same way that the Internet is addressed. A web-based information environment enables this flexible and rapid approach to targeting. Once a FAC had designated a target, an air controller working from either an E-3 AWACs or an E-2C Hawkeye was able to tap into the network and examine what assets armed with what particular weapons were available nearby and direct them to strike the target. As LGen. Keys noted, these grids could become increasingly precise themselves as the geographic information became increasingly precise itself: 'Think about not only having a Baghdad City grid, but a grid for each and every building in it. We should force an agreement on a uniform application and execution of joint live support coordination measures.'[23]

Here, it is possible to see the influence that computing power has on operations in specific. Geographic information system generated maps permit the integration of physical geographic parameters (location of building within a defined space) with human specific data (the type and function of building). These data can then be combined with equally complex information on not only the availability of armed assets flying in the area, but also their weapon loads. This permits air controllers to match the effect of the weapon against the nature of the target itself. The result is a highly tailored air strike that creates only the level of damage desired, reducing the possibility of collateral damage or fratricide. The availability of all this information in an easily accessible format, in turn, speeds the process of observation, orientation, and decision making.

The 'joint targeting cycle', which is at the heart of any air strike, is a rigorous process that potentially stands in the way of this rapid approach to targeting. In it, the commander's intent shapes 'target development, validation, naming, and prioritisation'. This in turn is followed by a capabilities analysis, the commander's decision and force assignment, mission planning, and finally, execution. While cumbersome, even in the case of 'time sensitive targets' (TST), the whole process must be respected as rushed application of force may result in fratricide, diversion of assets from higher priority missions, or collateral damage against politically sensitive targets, like the strike on the Chinese embassy in Belgrade during Operation Allied Force in 1999.[24] The process can be accelerated through rapid information exchange using web-enabled data that moves amongst the many decision makers that populate this process.

The growing importance of TST has required personnel working within CAOCs to game specific situations in advance, in the same way that naval commanders in the Persian Gulf gamed out interdiction and boarding operations in advance in order to deconflict ROEs.

> The strategy was in place so that a lot of decisions (about who and what to strike) were already made.... If it's this kind of target with this limit of collateral damage – if it pops up, then we're hitting it. You don't have to ask. That works for certain targets. (For example) if you knew that chemical weapons were released, and you knew that Scuds had them, and you found a Scud launcher, that would be a number one priority. Anyone on the ATO would go and kill it. On the other hand, if you had a leadership target in a car in a huge crowd, that's one that you wouldn't pre-decide. Someone would have to wring their hands over that one.[25]

Such an event specifically occurred during OIF in the air strike that was carried out at al Mansour when there were indications that Saddam Hussein was present at a meeting there. A 20-member targeting cell was able to develop the target information and move it through the process and relay it to the B-1B crew that carried out the mission within a cycle that reportedly only took 35 minutes. The crew itself only received the data 12 minutes before carrying out the attack.[26]

As one observer has concluded, in this accelerated process, the problems are only 20 per cent technical, the critical problems lie in the procedural elements of assessing and deciding on the information presented, especially under the pressure of time.[27] The implications are rather stark in terms of coalitions. Effectively the pressure of time and the classification of data means that many coalition partners may be shut out of this process. The need to release raw sensor data, or release it in a form which protects collection methods/technology, in order to arrive at a conclusion may effectively preclude any coalition participation in this process.

The US is clearly concerned about this aspect of operational integration. As the Chief of Air Staff, Gen. John Jumper noted in 2003:

> The buzzword for this decade is going to be integration. When you think of the basic principle of find fix track target engage and assess – this kill cycle that we talk about all the time – what's in every stage of that cycle? Command and control, intelligence, surveillance and reconnaissance. C2ISR. This is about timely decisions and knowing exactly where the target is well enough to do something about it.[28]

The solution to organisational stovepipes, according to Jumper, was to have to have the 'machines' talk to one another.[29] However, as the problem of TST shows, the technical aspect of this solution is only part of the problem. Machines will only share information that has already been categorised in some releaseable format: ultimately, foreign disclosure officers will have to decide what is releaseable to coalition partners.

NATO CAOCs are obviously different affairs than those operated by the US, yet even here, one might note specific distributions of nationalities in specific areas. One report noted that in the Ramstein NATO CAOC, the 'most internationally diverse' cell was the Peace Agreement Compliance Cell, which hosted two Canadians, one French officer, one British, and three Americans. Russians were present in the CAOC, but they worked out of the Air Mobility Cell, hardly the centrepiece of the unit.[30] Indeed, while Operation Allied Force was run from the CAOC at the Aviano Airbase, effectively, two ATOs were generated, one of which was US ONLY NOFORN in order to handle special targets and those dealt with by platforms like the F-117 and B-2. Further, NATO AWACS flying air control missions in support of Operation Allied Force were also not privy to such missions, sometimes only detecting them as the aircraft ingressed Serbian airspace.[31]

Such divisions caused by the compartmentalisation of information can even cause problems in purely unilateral missions.

> We had the cryptographic support group, the guys talking to NSA and the information war team on opposite sides of the wall from each other.... There were a lot of times they would run around the corner and say 'Do you know you can get this done right now instead of waiting for tomorrow's ATO?' Or sometimes we would go to them and say 'Something has changed (such as areas of jamming). Is that going to hurt what you're doing (with intelligence collection) tomorrow?'[32]

Such work-arounds are frequently commonplace within coalition operations, especially where high degrees of trust exist between partners, but they are always more inefficient than the vision of a seamless network outlined above. Indeed, even in a purely national AOC, the situation seems likely to get even more complex. Modern operations must frequently include interagency and NGO partners. As some have pointed out, the war fighting orientation of armed services means that lesser included tasks such as operations other than war are often considered last. Speaking of solving the technical problems of information sharing within an AOC, one study noted that priorities were 'air force first, then joint, then coalition, other government departments, NGO problems and first responders in that order'.[33] As such, technical advances and policy solutions to the barriers they raise contingently may outstrip the ability of external organisations to keep pace.

For all the functionality of web-based approaches to operations, such issues clearly are worrying. Allied and coalition forces have access to TBMCS, although with the caveat that such access depends on the 'particular coalition formed and the air war situation'.[34] For some partners, exceptionally, like the British, the level of access is clearly very high. The RAF operated closely with the USAF enforcing the southern no-fly zone over Iraq for 12 years. Air Marshal Torpy of the RAF noted that the operational drawdown for the RAF following the end of major combat operations in 2003 might place the level of interoperability with the USAF at risk: '... we must find new ways of providing this type of training for your young people ... (as) we may find ourselves in the next conflict not being able to work alongside the Americans so well.'[35] The example of NORAD is particularly instructive in this respect. The United States and Canada are two countries that share a common language, culture and increasingly, common economy and infrastructure. Despite everything that should unite these two countries, politically, there is a constant level of distrust between the two societies that if anything has increased in recent years. While this may reflect a passing moment associated with a highly unpopular presidential administration, examining the history of interaction within NORAD actually reveals declining cooperation between Canada and the United States, almost in an inverse relationship with the technical capabilities to cooperate.

Geographic destiny: North America as a unified battlespace

> Americans are destined to remain Canada's best friends whether they like it or not.[36]
>
> Robert Thompson

The management of North American security is conducted under the concept of bi-nationalism, a condition that emerged in the growing uncertainty of the late 1930s and the threat posed by Germany. It prospered during the Soviet challenge of the Cold War, and has been to some extent, renewed by the ongoing War on Terror. For many years, bi-nationalism was the source of unprecedented military cooperation between Canada and the United States in the realm of air defence, even as it provoked wider strategic concerns on both sides of the border about the inherent limitations it placed on the sovereign independence of both states. Clearly, the conjunction of North American geography with its vast undefendable borders, the close cultural linkages between each society, and the enormous power differentials between Canada and the US has defined this political relationship between the two neighbours. Geography and culture ensure that neither nation can unilaterally move away from the tight embrace they share, although as the sole variable, the power relationship between each country provides considerable dynamism that has a key impact on the operational arrangements the two militaries must work within. NORAD emerged as an operational solution to a bi-national security problem. Strategic relations between the two nations have since worked to scale back military cooperation.

Speaking to officers who have long worked within the NORAD framework, one often hears the notion of the NORAD habit, a style of working in which the strategic, operational, and tactical concerns of the moment are filtered through a mindset that considers how they will impact on the bi-national approach of the organisation. While a critical aspect of the relationship, it is a necessary but insufficient component of it. It is supplemented by a further 850 agreements and other documents relating to North American security.[37] Like the British constitution, the Canada–US defence relationship is a largely unwritten document, codified in terms of practice.[38]

NORAD is part of a rare, perhaps singular, strategic relationship between nations. The 'bi-national' as opposed to bi-lateral relationship is a 'dense and extensive defence architecture based on partnership and largely expressed through bi-national institutions', according to one former American co-chair of the Permanent Joint Board of Defense, one of those institutions.[39] As Colin Gray sought to describe it, it is the notion that the Canada/US border is an 'irrelevant and academic legal construct' in terms of the defence problem shared by both nations, where each nation views its security in fundamentally similar terms.[40] One former NORAD Deputy Chief of Staff for Operations, who is always a Canadian officer, described it thus:

> It's the Canadians and Americans *together*, doing the job *together*, and *together* providing information to Canada and the US at the same time. So it's *totally* integrated, seamlessly and totally integrated and authority is delegated to Canadians which you don't see the Americans doing anywhere else.[41]

Another member of the Bi-national Planning Group (BPG) explained it similarly: 'Bi-national, to me, means you put Americans and Canadians together to do something.'[42] LGen. Eric Findley, a former Deputy Commander of NORAD explained it poetically as akin to 'being married. And bi-lateral is less than married. It's out there in the common law, let's live together realm.... A Canadian can tell American officers ... here's what to do.'[43] Finally, the present Deputy Commander agrees with his predecessor on the close relationship between the Canadian and American military within NORAD. He pointed out that in the bi-national construct, Canada and the US cooperate closely over decision making, whereas in a bi-lateral relationship, Americans could easily make unilateral decisions, only informing Canada afterwards of the fact. Aircraft assigned to NORAD can technically be considered neither Canadian nor American, but NORAD assets.[44]

This unique command relationship between two countries evolved slowly, beginning in the interwar period and the growing tension that accompanied the rise of Nazi Germany. In August 1938, President Franklin D. Roosevelt delivered an address at Queen's University in Kingston Ontario promising that Americans would 'not stand idly by' were Canadian security to be threatened by an outside power. Shortly thereafter, the Canadian Prime Minister, William Lyons Mackenzie-King issued guarantees that Canada would ensure sufficient security measures so that it did not become a threat to the United States.[45] This series of declarations, which effectively initiated the modern period of Canada/US defence cooperation, is the first political recognition of the indivisibility of North American security. The declaration would be followed by a series of political institutions including the already mentioned Permanent Joint Board of Defence (PJBD) in 1940 in which each country participated as equals despite the enormous disparities of military power between them. These intensified in the immediate post-war environment. The uncovering of a Soviet spy network within Canada by the 1945 defection of Igor Gouzenko provoked heightened concerns; according to Mackenzie-King, 'Canada could not do what was necessary to defend itself', however, in his view, it was important not to replace the abandoned and unsatisfactory bi-lateral security relationship with the British for a similar one with the United States.[46] Rather, the

PJBD developed a Military Cooperation Committee in 1946, a further bi-national institution, which shortly thereafter published a 'Joint Canadian–United States Basic Security Plan' that outlined a common perspective on the threat to North America.[47]

Within this evolving bi-national partnership, consideration of the air threat to the continent began to be considered between both air forces. In 1947, General Earl E. Partridge (USAF), CINC Continental Air Defense, observed that 'the air defense of Canada and the US is one problem and that both countries will react automatically and in unison against any attack on the North American continent'. This was fully appreciated by his Canadian counterpart, Air Chief Marshal C. Roy Slemon, then Chief of Air Staff RCAF, however in a letter to General Charles Foulkes, Chairman of the Chiefs of Staff Committee, he doubted that Canadian politicians would agree.[48]

In a tour de force that has rarely been equalled in the history of Canadian defence policy, the Canadian military 'stampeded' the government into agreeing to a joint operational command shared between Canada and the United States that would oversee the operations of North American air defence. This involved a series of backroom deals between the RCAF, the USAF, and ultimately the Joint Chiefs of Staff, where the latter were quite literally used as cut outs in order to achieve the operational plans both the RCAF and the USAF saw as necessary to defend the continent from any Soviet air attack.

While Canadian air force officers implicitly understood the strategic logic of air defence cooperation with the US, Canadian politicians were cool to the idea and stalled any progress. Within the United States, scepticism was confined to the Joint Chiefs of Staff, who doubted that Canada could be relied upon to act in a crisis and feared the meddling of Canadian politicians and diplomats. Ultimately, the RCAF suggested to the Americans that they write a proposal arguing for a joint command to oversee North American air defence. The JCS was convinced by the intervention of the Air Force Chief of Staff, General Nathan Twining, although they noted that any organisation should aim solely for 'operational integration' as 'a combined Canada–US command is probably not acceptable to the Canadians at this time and should not be proposed'. The proposal suggested extending only 'operational control' over forces assigned to the organisation.[49] As General Foulkes remarked: 'There were no boundaries upstairs, and the most direct air routes to the US major targets were through Canada. Therefore, air defence was to be a joint effort from the start.'[50]

This close operational relationship would come back repeatedly to haunt the organisation, highlighting the problem of failing to take into account the political implications of any military operational arrangement, no matter how professionally cordial. Still, the idea of a single integrated battlespace has rarely been challenged in its fundamental nature. Despite the political difficulties that affect Canada/US relations, they are never so deep as for one to view the other as a fundamental threat to its very existence. Even after the attacks of 9/11, American concern over the perceived lack of Canadian security was framed within the context of foreign elements infiltrating across the Canada/US border.

In some respects, 9/11 strengthened the concept of an integrated continental approach to North American security. Speaking in 2006, General Rick Hillier, Canadian Chief of Defence Staff remarked that in the post-9/11 world, Canada and the US had to deal 'not with the bear, but a ball of snakes; not with ICBMs, but every object in the sky; not with threats outside of North America, but within it'. Such a strategic problem would require much greater levels of cooperation between the two nations, extending beyond strict military-to-military planning, and include a whole raft of government agencies.[51] In this environment, the BPG observed that continental defence would require 'the orchestration of all Canadian and US elements of national power', diplomatic, informational, military, and economic.[52] General Ralph Eberhart, Commander of NORAD, noted in February 2004 that:

> My intuition is that we need to take NORAD to the next level.... For sure, we need to include some kind of maritime piece ... and probably some kind of civil support.... We should have the ability so if one nation asks, the other is ready to respond on the shelf, ready to go as opposed to working through the bureaucracy.[53]

Operational/strategic relations

Within NORAD, information sharing is surprisingly open. The diplomatic notes governing information exchange between Canada and the US were only established in 1962, perhaps significantly, following the Cuban Missile Crisis. However, the notes only mention the protection of information and how shared material can be used by either party. They say nothing on the rules governing sharing itself.[54] Nevertheless, from the very first days of NORAD, the USAF went to considerable lengths to ensure that information was shared as widely as possible. According to Air Marshal Slemon, NORAD's first Deputy Commander,

> (Partridge) said 'Roy, I'm supposed to be the Commander in Chief of NORAD and you are supposed to be the Deputy Commander in Chief. When I go on a trip ... you have the responsibility and the authority. I can't go away on these trips and have any peace of mind because you don't know what the hell goes on with regards to the (nuclear) weapons. So as of this minute, you are privy to all that is necessary with regard to the nuclear weapons.' He never referred to headquarters or anyone. He made the decision right then and there and the word was passed on. He was never rebuked by his superiors and the guy took it on. It could have cost him his commission because the security on those weapons is top.[55]

More recently, however, General Findley and others in NORAD remarked that the twin decisions on the part of Canada to not participate in either Iraq operations or in ballistic missile defence (BMD) had precipitated some restrictions on information sharing. Findley noted:

> Do I have any empirical data that things have changed since the decision on missile defence, or the decision not to participate in Iraq? No, because they don't say or not.... But it is quieter. Or there is less information than there was.

One other senior Canadian NORAD official noted that the February 2005 decision not to participate in BMD had a dramatic effect on efforts to enhance releaseability policies on missile defence information.[56]

Still, in the autumn of 2005 members of the BPG could claim with pride the significant efforts that had been made by both Canada and the US to link each nation's secret networks, the US SIPRNET and the Canadian TITAN system through the already mentioned GRIFFIN network in a link known as SIPRNET REL A. The BPG report included recommendations for a 'write to release' approach in the hopes of reducing the problems that accompanied NOFORN documents.[57] NORAD, in conjunction with Strategic Command continued to roll out the Combatant Commander's Integrated Command and Control System (CCIC2S), which would provide a 'common infrastructure to share and maintain information across the air, space and missile warning areas'. The system would permit the fusion of information from NORAD's regional AOCs and air defence sectors to provide a common operational picture for NORAD, Strategic Command, and Northern Command. CCIC2S includes segments for air defence, missile warning/defence, space command and control, and core services infrastructure.[58] Further, they also noted the presence of three Canadians within the Combined Intelligence Fusion Center, an organisation then feeding information to both NORAD and Northern Command.[59]

A further indication of the high level of trust between each nation's air forces was demonstrated in the autumn of 2007. Following the crash of an Air National Guard F-15C in Missouri, the USAF's fleet of F-15s were grounded worldwide due to fears of structural fatigue in the airframe.[60] The jets remained grounded into 2008; in January, General John Corley of Air Combat Command noted that they would only return to the air on a 'plane by plane' basis after passing a series of structural inspections.[61] In Alaska, where F-15s perform sovereignty patrol duties for NORAD, CF-18s from CFB Bagotville in Quebec stood in for the grounded jets until they could return to duty. Operations included intercepts on Russian Bear Tu-95H bombers, which had only recently begun challenging North American air space as they had throughout the Cold War.[62]

In terms of information exchange, the organisation seemed to have weathered the storms accompanying the missile defence decision. Within the Missile Warning Center, a Command Center within the Cheyenne Mountain complex, commanded on a rotating basis between Canadian and American officers and with a substantial bi-national staff, information is received and passed along to American missile defence organisations.

> (The Missile Defence Officer) is an American Only position, but of course, he is sitting in the same room with both Canadians and Americans. So what

we've done is, essentially, position screens so that they are not immediately visible.... What we've done is allow Canadian access to some of the conference feed on the telephone without all the conference feed. We've put limitations on what Canadians can say and do on particular conferences and make sure that they don't cross the line into active participation into missile defence. So, in the end what they call 'incidental exposure' to missile defence operations will occur, because that is the only way you can keep Canadians in the Mountain, but we are not heavily, directly involved in the decision making.[63]

By 2008, with relations substantially improved between the two countries, LGen. Bouchard was able to assure a group of Canadian officers convened for an update on Canada–US relations that information sharing was substantial and effective at his level. He noted that he had been made privy to information on a variety of classified programmes that had implications for North American air defence and was also privy to information on BMD programmes and foreign missile intelligence.[64]

This close relationship, however, frequently provokes alarm amongst some Canadians, especially considering how the organisation was created through the professional collusion between the RCAF and the USAF.[65] NORAD is only delegated Operational Control of forces assigned to it, not full command. Even today, such distinctions nonetheless continue to confuse those not familiar with them.[66] In an environment of even slight strategic distrust, much hay can be made of them. Two years after the signing of the NORAD agreement in 1958, Canadian journalist James Minifie alleged in his book *Peacemaker or Powdermonkey* that Canada was 'sucked in' by the 'brassy intrigue' of the USAF and had subordinated its independence to the 'one over-all boss in the USAF'.[67] Forty-two years later, the same arguments were still being recycled within Canada. One prominent Canadian academic has recently sought to argue that American operational control of Canadian Forces would place Canada's commitment to human security at risk by conceding a veto power over CF operations. Michael Byers argues that what is at issue is not Canada's legal sovereignty but rather the practical limits on that legal right: 'its ability to freely make choices at the international level'. Such concerns 'cannot be overcome by the technical distinctions between command and operational control'.[68] The history of NORAD demonstrates such fears are clearly mislaid,[69] yet nevertheless, even after 50 years of relatively problem-free interaction, they continue to contribute to the climate of suspicion between the two countries.

Still, as ludicrous as Minifie's argument was given the active role played by the RCAF in shaping American military opinion, his point goes to the problem of operational agreements that are not well grounded in the politico-strategic nature of interstate relations. Early in 1957, much as the Joint Chiefs of Staff (JCS) feared, the cosy operational agreement between the RCAF and the USAF was challenged by both Canadian diplomats, who had been long working to facilitate a process of consultation between Canada and the US in the event of a war crisis, and Canadian politicians, who feared that Canada had given up its right to declare war. On the US side, General Partridge increasingly feared that the Canadian government might withdraw its forces in a crisis, as 'automatic' involvement had not been formally agreed to. Of course, the speed with which an attack could be mounted, the very thing which had spawned the close cooperation in the first place, mitigated against formal consultation in a crisis. Alerts would have to be announced quickly in order to get as many aircraft into the sky as possible. Still, an alert in the midst of the crisis might also be interpreted as a signal that the US was about to strike, potentially committing Canada to war before one began. Plainly, Canadians would want to fully consider the implications of their actions, based on the fullest amount of information that could be made available at the time.[70]

Here an irresolvable strategic conundrum confronted American policy. As the leader of the Western free democracies, the US had extended security to its allies in the form of its nuclear retaliatory arsenal. Canada was deeply embedded in this arrangement; NORAD was formed as a defensive bulwark to that arsenal, ensuring that it could survive a nuclear strike. However, European NATO partners feared first that the US would seek to re-fight the Second World War on their soil, shielding America from the damage of major industrialised war. Second, they also feared that in a pinch, the US would abandon them to their fate. The challenge here was to guarantee 'coupling' of America's strategic arsenal to the tactical battle that might be fought in Europe, in effect that the US was willing to fight a nuclear war on their behalf. Of course, this entailed being ready with operational plans to actually use nuclear weapons to fight a war.

Canadians, on the other hand, faced different strategic concerns. Less concerned with a Soviet invasion of the North American continent, they worried about being dragged into wars not of their own making and over which they had less say. As such, Canadian policy was philosophically opposed to any plans which were oriented towards the use of nuclear weapons as operationally useful tools of war. Enunciated by the Trudeau government under the concept of Mutual Stable Deterrence, the only use for nuclear weapons, from the Canadian perspective, was to create mutual fear between the superpowers, ensuring war would not break out in any case.[71] As the Defence White paper *Defence in the 70's* put it:

> The only direct external threat to Canadian national security today is that of a large scale nuclear attack on North America. So long as a stable strategic balance exists, the deliberate initiation of a nuclear war between the USSR and the US is highly improbable; this constitutes mutual deterrence … therefore, Canada must do what it can to ensure the continued effectiveness of the deterrent system … From a potential enemy's point of view, however, North America can only be seen as one set of targets…. The government concluded in its defence review that cooperation with the US in North American defence will remain essential *so long as our joint security depends on stability in the strategic balance.*[72]

Ultimately, in a secret annex to the NORAD agreement that was negotiated between the two countries in 1958, each agreed that

> In a situation in which either Government concludes that alert measures are necessary or desirable, both in the USA and Canada, the two Governments agree to consult through the diplomatic channel and through the respective Chiefs of Staff of the two countries.... If either Government is impelled by the time factor to take alert measures before initiating consultation, it agrees to immediately inform the other Government of the action taken and to consult with the other Government as soon as possible.[73]

The agreement was put to the test in 1961 during the Cuban Missile Crisis, when a request to place Canadian Forces on alert was refused by the Prime Minister Diefenbaker for two days because he felt that he hadn't been sufficiently consulted by the Kennedy administration. Fittingly for the nature of the agreement's origins, the Defence Minister secretly ordered the Canadian Forces to go on alert quietly. Still, the fears of both Canadian politicians and American military commanders had been confirmed by the crisis. Colin Gray suggests that in 'the context of a small to middle power–superpower relationship', the hope for such a full and open consultation was 'forlorn'.[74]

Managing the differences in strategic interests of its allies distinguishes American approaches to NORAD from Canada's.[75] America was also concerned with its own sovereign capabilities as well, specifically to engage in operations it saw as central to its national security but on which Canada might have objections. Throughout NORAD's history, its commander was always 'double-hatted' as commander of a US-only command. NORAD had effectively evolved out of the USAF's Continental Air Defence Command, or CONAD. While Partridge had considered that CONAD should be disbanded, the JCS objected to ceding total control for air defence to the bi-national NORAD. As the bomber threat to North America receded, the importance of CONAD faded. At the same time, as the missile threat grew, and as American space assets grew, new Commands developed. Ultimately CONAD was disestablished and Space Command was created and 'twinned' with NORAD. Here again was a US-only organisation through which American military operations could be effected without Canadian participation if necessary. In 1985, US Element NORAD, staffed entirely with those American officers already in NORAD postings, was also created as a further back door to conduct purely national air defence operations without the participation of Canadian Forces.[76]

Space and missile defence

The American scholar of NORAD, Joseph Jockel has observed that 'the US would trust the military of no other ally in the assessor position, not even the British. It is striking that the US still feels this way, given how little Canada contributes to North American aerospace defence.'[77] Canadian officers work in the heart of the Command, as assessors evaluating the data that flow into the Missile Warning Center and making recommendations for the Director of Operations, a shared Canadian/American position, and ultimately for the Commander and his Deputy, a process known as Integrated Tactical Warning and Attack Assessment (ITWAA). This information flows in from a variety of sensors, many of them located in outer space. Canada has had a changing relationship with American space activities.

As missiles gradually replaced bombers in the threat to the North American nuclear retaliatory response, space surveillance became an integrated mission within NORAD as well. Here, the absence of assets that Canada could deploy initially restricted its participation in this sphere. The Ballistic Missile Early Warning System (BMEWS) radars were situated in Alaska, Britain, and Greenland, outside of Canada. Two Baker Nunn cameras, used for tracking satellites, were placed in Alberta and New Brunswick, however. While Canadians worked within the Missile Warning Center, they had no positions within the Space Defense Center. Even at this early date, missile defence raised issues with regards to continental cooperation. The Commander of NORAD, through his role first as Commander Continental Air Defense Command (CONAD) and later as Commander Air Defense Command (ADCOM), had operational command of the Safeguard missile system. The Deputy Commander at the time, LGen. Lane, was able to force the issue that Canadians would require the information to operate effectively in the headquarters and two positions were created for Canadians in the Space Defense Center.[78]

Still, the role and participation of Canada in military space activities would remain controversial in both the US and Canada, irrespective of the desires of the Canadian military. The creation of Space Command in 1985 recognised the growing importance of space for the American military. Unlike the previous 'twin' command, ADCOM, Space Command had no role in air defence and thus no natural link in which Canadians could participate. Indeed, Jockel goes so far as to argue that the US either 'did not want to, or they were explicitly forbidden to' cooperate with Canadians in these affairs, although they had gone to extraordinary lengths to gain Canadian permission to approve the double-hatting of the NORAD Commander.[79]

The bi-national nature of NORAD, the integrated role Canadians played in the ITWAA process, and the fact that many Space Command personnel, at least in senior levels, were also NORAD personnel seems to have eroded these firm institutional lines, so much so that one Deputy Commander NORAD could admonish his staff that 'Canadians are not in Colorado Springs to infiltrate Spacecom!'[80] However, once again, the operational agreements that the Canadian military could effectively arrange with its American counterparts ran up hard against the strategic positions that both countries were pursuing internationally. That Canadian military personnel felt professionally frustrated is no surprise as the Canadian position was extraordinarily nuanced, some might uncharitably call 'contradictory'. Canada was committed to Western defence, and thus to the existence of a nuclear retaliatory response as the basis for strategic deterrence.

However, as a small military power Canada was instinctively distrustful of strategies which sought to employ those same weapons in any case other than an all-out response to a surprise attack, fearing being drawn into a war against its will.[81]

The relationship between NORAD as a warning and air defence coordination centre was carefully delineated from offensive/retaliatory functions that the Strategic Air Command might pursue, or from missile defence roles themselves. The distinctions here were extremely fine as information passed through NORAD ultimately permitted SAC, Space Command, and at present Strategic Command to execute warfighting missions in which Canada was not involved, nor wished to be. Politics might keep these missions separate, however, technology linked them together and ultimately placed Canadian officers in uncomfortable positions in between their operational partners and their own government.[82] Canadian military officers feared that the diplomatic policy rejecting any Canadian participation in active missile defence would ultimately marginalise the Command to the point where it might become entirely irrelevant.[83]

The rancour that had accompanied Canada's decision not to participate in the invasion of Iraq concerned many that the Canadian–US defence relationship was falling apart. Indeed, both Joel Sokolsky and Joseph Jockel have been warning for some time that this was an inevitable process.[84] Following the retirement of Prime Minister Jean Chretien in 2004, his successor Paul Martin made repairing Canada–US relations a central part of his government's policy. Part of this was a seeming about turn on Canada's long-term objection to missile defence. The end of the Cold War stand off between Russia and the US, the latter's withdrawal from the ABM treaty, and the absence of any resultant arms race seemed to have removed the logical underpinnings of Canada's objection according to its own policy of mutual stable deterrence. Further, the obvious willingness of the US to pursue its own interests together with the fear that this might undermine the cooperation within NORAD played a critical role in this shift.[85] Even then, no real significant shift was anticipated with one report anticipating that Canada would keep a very traditional role in any missile defence system, confined perhaps to an assessment and warning role.[86]

But just as the close operational relations between air forces had strategic implications for the Canada–US relations, the connections between the strategic and the operational level in reverse were all the more direct. Despite the growing indications that Canada would sign on for a role in any continental missile defence system, the Canadian government made an abrupt U-turn and halted further discussion. It assessed that a positive decision would have a critical impact on a by-election being held in the spring of 2005 in the province of Quebec, the area most opposed to closer defence relations with the US, as well as the growing levels of opposition to such an accord within the ruling Liberal party caucus. As with the decision not to participate in Iraq, the announcement of the decision was mishandled, reportedly Prime Minister Martin abjuring from informing the President himself at a NATO meeting, relying on his Foreign Minister to pass a note to Secretary of State Condoleezza Rice. The US reaction was blunt:

> For the record, the US is disappointed by Canada's decision on missile defense. We wish it had been otherwise but we equally accept that it was always Canada's decision to make. We will move on with our important cooperation in the defense of North America.[87]

As noted above, the impact of the decision was subtle and some Canadian officials noted that the implications could actually be rather overstated.[88] Further, even before the decision, James Fergusson pointed out the obvious solution that Canada could participate without participating completely by restricting the execution of missile firing to Americans, an arrangement that effectively functions today in any case, as discussed above.[89] Nevertheless, close observers of NORAD have speculated that the decision ultimately will have 'profound implications' for NORAD as new missile defence systems come online that have few or no connections to NORAD itself. Ultimately such developments could undermine the role of NORAD in the ITWAA process itself.[90] What was clear was that with the movement of Space Command's functions to Strategic Command in 2002, a no decision meant there now would be no connection of Canada to American space developments. In the words of one observer, 'Canada is no longer needed and will likely no longer be petitioned.'[91] It would be easy to conclude that Canada will be unilaterally shut out of space, except that each government has mutually arrived at the same solution: committed to its doctrine of Mutual Stable Deterrence, Canada has placed clear boundaries on its cooperation with the US, whereas the latter continues to pursue its global military engagement through investment in space technologies, a mission which is easily separated from the imperatives of bi-nationalism.

9/11 and the realignment of North American security

The operational relationship between Canada and the US had been long undergoing a slow regionalisation culminating in the 1980s reorganisation of the NORAD Region and Sector boundaries (see Figures 5.1 and 5.2). NORAD had taken a continentally integrated command structure and broken it up into three separate regions, one of which was devoted exclusively to Canada. While this was obviously reassuring to those fearing American control of Canadian forces, the realignment broke up the integrated structure of the organisation along national barriers, arguably reducing its bi-national character and impeding the flow of information.[92]

The events of 9/11 realigned considerations of North American security in several important ways. In some ways, it reinforced the notion of the continent as a single battlespace in ways that had perhaps been slowly fading given the collapse of the Soviet threat. In the 1990s, NORAD had become a bit of a backwater and there were even thoughts that it might be closed altogether. However, security now became a much wider concept than ever envisioned under the Command and new missions began to be considered in addition to that of traditional air defence and missile warning. These would all involve new partners

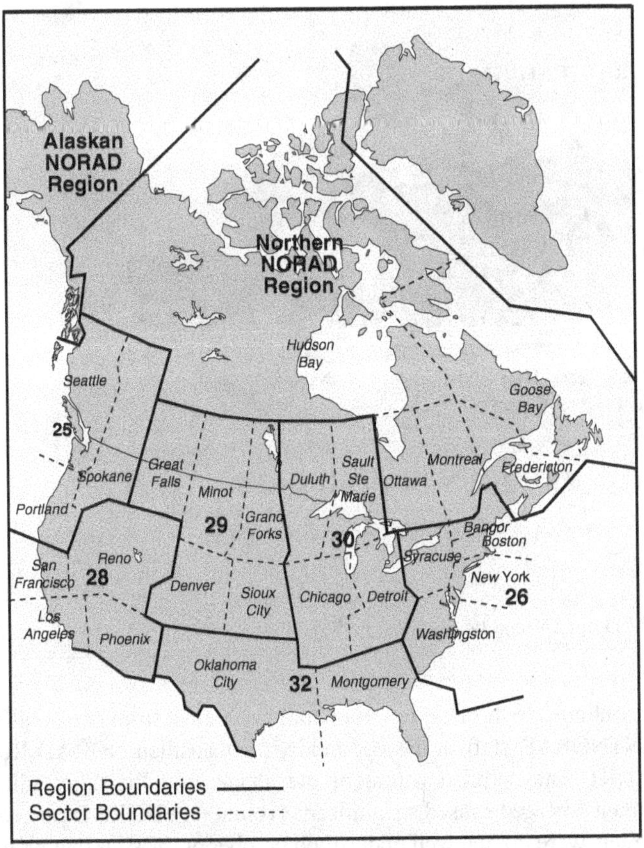

Figure 5.1 NORAD operational boundaries, *c.*1961.

and, to some, a troublesome set of new organisations both within each country's military as well as the wider government. Indeed, just as the continent seemed to be unified in terms of geographic appreciations of security, these changes threaten to rein in cooperation between Canada and the US still further.

The consequences of 9/11 produced a series of surprises for the organisation. Most obviously was that the strikes emerged from within the continent, rather than flying across the poles or otherwise crossing the maritime boundaries of the continent, something entirely unanticipated by the command. Furthermore, despite years of attempting to distance NORAD from NATO,[93] the first articulation of Article Five of the Washington Treaty calling for all partners to come to the aid of one under attack followed the strikes on New York and Washington DC. NATO E-3 AWACS deployed to Tinker Air Force Base in Nebraska for a 220-day deployment in support of the NORAD Operation Noble Eagle.[94]

Figure 5.2 NORAD operational boundaries, *c.*1985.

Noble Eagle continues to this day, run as operations distinct from one another by the USELEMENTNORAD staff in the US and by the Canadian NORAD Region in Canada (CANR). The surprise nature of the attack, and the non-traditional vector from which it emerged caused a significant evolution in NORAD procedures and policies. Targets were not the well understood bomber or missiles that NORAD had long prepared for, but civilian aircraft. New tactics and scenarios had to be prepared for, indeed, a whole new regime governing Rules of Engagement had to be developed rapidly on the day of the attacks itself. General Findley notes 'We had to put some Band Aids … in place.… The Band Aids are gone now with some permanent solutions in place.'[95] The clear differences in Canadian and American rules of engagement make this command distinction necessary: while US commanders have authority to order the destruction of civilian aircraft in US airspace, the Canadian Deputy Commander does not. Nor do Canadian commanders have similar authority in Canadian airspace. That decision is still reserved by the Canadian government.[96]

Organisationally, as well as tactically, NORAD was caught unprepared. While as the Director of Operations on the day of the attack, then MGen. Findley ordered all NORAD assigned assets in North America to 'generate generate generate' in order to place as many air defence assets in the sky as possible, the 'battle' was run from the Northeast Air Defense Sector Command in Rome New York, rather than from the HQ in Colorado Springs.[97] Further, the organisation was unable to communicate effectively with the Federal Aviation

Authority (FAA) who first noted the existence of the threat as air traffic controllers lost control of the flights. As the subsequent 9/11 Commission found, the absence of direct links between the FAA and NORAD hampered the transmission of information.[98] While some improvisation occurred, FAA officials had difficulty in joining teleconferences with NORAD controllers directing the response.[99] Despite improving linkages with the FAA, including bringing a representative within the NORAD Command Center, as recently as 2005, NORAD continued to experience difficulties in passing data from it to the FAA. *Inside the Air Force* reports that on 12 May 2005, NORAD was unable to pass electronic data on a plane approaching the Washington DC Air Defense Identification Zone to FAA staff seated in Cheyenne Mountain itself 'because the NORAD files could not be uploaded on to the FAA run networks'.[100]

General Gene Renuart summed up the situation succinctly noting that NORAD now faced the challenge of dealing with 'rogue elements' as opposed to the more structured challenges posed by the Soviets during the Cold War. In addition to the problems posed by rapidly evolving technologies such as UAVs and cruise missiles, new issues like the impact of climate change on the Arctic region might spark security challenges from unpredictable directions.[101] The seamless network aimed for during the Cold War in the form of a comprehensive system of radars would now require a far more complicated series of information inputs, ones that would firmly reinforce the continuing relevancy of Canada to US security. Air defence under the new threat conditions was impractical without Canadian civil aviation data. Further, maritime security must now be considered, as threats could easily emerge from coastal and riverine areas; finally the growing integration of North American infrastructure meant greater levels of coordination between governments themselves – North American security was much larger than simply standard military preparations.[102]

Into this new complex environment, new organisations began to emerge. On the civil side, the Bush Administration created the Department of Homeland Security, and on the military side, Northern Command was created to oversee the security of the Northern hemisphere. These were complemented in Canada with similar organisations, Canada Command, a military command that would oversee all domestic operations, save those overseen by NORAD itself, and Public Safety and Emergency Preparedness Canada (PSEPC). Along with these new departments, older actors such as the FBI, the Royal Canadian Mounted Police, customs and immigration departments on both sides of the border, and an entire raft of actors at state, provincial, and municipal levels needed to be considered as partners. Many of these organisations had cooperated across the border with their counterparts in either country, however, NORAD was the only bi-national member of this new constellation, and one focused on a narrow mission.

The Bi-national Planning Group (BPG) was formed in December 2002, largely in response to the emergence of Northern Command, in order to consider how the new security environment would affect the relationship between Canada and the US, particularly in a period defined by a strong unilateralist tendency within the US and atrophied Canadian military capabilities.[103] Reporting four years later, the BPG advocated among other things, a series of new missions that would be adopted by NORAD, including maritime surveillance and cooperative planning for consequence management of natural disasters or terrorist strikes.

Given the long history of maritime cooperation between the Canadian and American navies, and the common vision of maritime security within each country, the new maritime surveillance mission seemed relatively unproblematic.[104] Furthermore, such a mission fit well within the evolving concepts of Maritime Domain Awareness (MDA) and the Proliferation Security Initiative (PSI), which both envisioned a broad network of navies and maritime security organisations cooperating in order to enhance regional security. However, the level of cooperation entailed in these efforts was clearly more broad and diffuse than the close cooperation that had characterised NORAD throughout its history. Taken within an MDA or PSI framework, maritime surveillance might actually require less cooperation and less information exchange than had been practised by NORAD! Obviously, this made such initiatives easier to implement, but they could hardly be characterised as an advance.[105]

Canadian nationalists were predictably concerned that the 'Civil Assistance Plan' (CAP), a series of plans to allow the armies of either nation to operate on either side of the border in the event of an emergency, were simply one more example of disguised attempts on the part of the American military to act in defiance of Canadian sovereignty.[106] In reality, the CAP sought to enhance the sovereignty of each nation. It would examine the differences in ROE that both nations would employ in the event of particular emergencies, and attempt to resolve them in advance. Further, it considered which niche military specialties would be desirable to deploy in specific emergency scenarios, like medics, engineers, and CBN specialists.[107] Through improved information sharing, trust between each partner could be enhanced in the event of a serious crisis requiring the closure of the border. The fear within the BPG was that the intense media attention that would accompany any disaster might have contingent effects complicating the management of a crisis. As each organisation would know how the other side was responding in advance, the increased trust would reduce the amount of time such events would compromise border security.[108] Ultimately, some saw the CAP as no different than sending a disaster response team in response to some natural calamity, as Canada had had opportunity to do twice in recent history, following Hurricane Andrew in 1992 in Florida, and after Hurricane Katrina in 2005.[109]

Despite these optimistic plans, however, the BPG was regarded by some as bureaucratically isolated and there was a lack of an enthusiastic reception on the part of the US military to its final response.[110] Bureaucratically, the establishment of Northern Command seemed to pose real threats to Canada–US cooperation as it had come to be characterised by NORAD, one observer noting that Northern Command's establishment had 'trumped a long standing diplomatic agreement, rendering it nearly, if not completely irrelevant'.[111] Others too pointed to the fact that Northern Command was more a competitor to NORAD than a 'twin' as Space Command and CONAD had been. For Northern Command, Canada Command was the natural partner, and as a unified regional Command, it was legally delegated to promote all 'theatre security cooperation'. Combined with the threats to ITWAA by the transfer of space assets to Strategic Command, the possibility that NORAD might be downgraded to a mere joint task force overseeing a much reduced air defence mission, itself increasingly run along national rather than bi-national lines, suggested a shrinking of cooperation between the allies.[112]

As the organisation completes its first 50 years of existence, it confronts new challenges to its very nature. Almost immediately following the formalisation of the agreement in 1958, it had to begin dealing with changes to both the mission and environment. In all, the organisation managed to rise above the challenges posed by the growing threats posed by ICBMs and the development of space capabilities. Despite the limited or absent investments on the part of Canada in response to these issues, NORAD continued to function very much like its founders envisaged. Some have argued that the new asymmetric threats that confront North America share similar characteristics as the classic ones that confronted NORAD throughout its history. Threats would be largely surprises and a speedy and combined response would still be required from both partners.[113] However, clearly, the nature of the new threats and the new partners were raising serious issues with respect to the habits of cooperation that had been developed over the past 50 years. Even the most enthusiastic observers of bi-national relations had to admit that there were 'complex doctrinal, legal and information sharing problems within each country and between them'.[114]

As the BPG observed, in the new complex environment of continental security where threats could emerge from unanticipated vectors, all agencies play a role and need to be incorporated. As they pointed out in a term that has now been adopted outside of NORAD itself, there is requirement for a 'need to share' policy as opposed to a 'need to know'. Here, the lack of political agreements to shape information sharing between and amongst new partners, together with long standing organisational resistance to sharing by groups unused to such approaches, means that there are immediate impediments to sharing.[115]

It is precisely this aspect which raises real concern on the long-term viability of a truly bi-national approach to North American security. Currently, NORAD's operational picture is developed by information flowing in from 25 separate feeds, nearly half of which are non-Defense Department ones. Northern Command receives information from over 700 state and local organisations and has 'strong working relationships' with 70–80 of them.[116] Many of the sources are not releaseable to Canada. The problem stems from the new bureaucratic organisations being added to the mix – while the US military might arrive at work-arounds to share information with its Canadian partners, there are serious difficulties in sharing intelligence from the security and law enforcement communities, especially in a lateral fashion.[117] Domestic stovepipes, difficult enough to deal with within a purely national context, as the 9/11 Commission demonstrated vividly, become all the more problematic in the bi-national environment of NORAD.

The information network that is NORAD poses the real problem in this regard. Within the law enforcement, intelligence, and other security-related communities, transborder information sharing sometimes is easier than cross governmental information sharing. Thus, the RCMP may have excellent relationships with the FBI, but that good relationship does not extend so far as to permit FBI-derived information to be shared with Canada Command. NORAD officers all remarked on a need for a cultural shift on the part of new agencies working in this realm. This in itself will be highly problematic, irrespective of the clear need for greater cooperation. As LGen. Findley remarked

> It's a hard leap, both here in the United States and probably the same in Canada for law enforcement to get their arms around the requirement to share things. Here in the United States and in Canada, people are loath to share information because they don't want to jeopardize an on-going investigation or the jurisprudence that is taking place.[118]

As another NORAD official noted, even an acknowledgement on the need to share at upper levels of command did not translate into any 'sense of empowerment on the part of middle ranking and lower ranks. So people are worried that they don't have the authority to release.'[119] Finally, clearly the cross-federal nature of information sharing in this environment also compromises effective information sharing. Irrespective of Presidential, DOD, or Pentagon orders to make information more available, neither State nor local level authorities are explicitly beholden to comply with such requests.

Similar problems confront the evolving maritime mission that NORAD has undertaken. The navies of both Canada and the United States are reportedly institutionally opposed to the overall mission, fearing that it might compromise their ability to deploy overseas.[120] In any case, any such cooperation was clearly going to be highly limited. No forces have been assigned to NORAD to accomplish this mission, unlike the air defence mission. Thus, the organisation will gather information 'where it can'.[121] The best that can be expected with this approach is a general situational awareness, not the bi-national unified response that characterised the air defence efforts of the Cold War.

Conclusion

The BPG noted 'although both nations have stated that transformation and network centric operations are their strategic goals, these concepts have not been fully implemented in a bi-national environment. Political direction must occur to affect that change.'[122] This is in itself a strong indication of the likely trend in terms of bi-national cooperation between Canada and the United States. Irrespective of the excellent operational cooperation between their respective militaries, the level of strategic cooperation seems to be in a state of decline. This is directly in opposition to the level of integration between their two societies, which have shown growing levels of integration in a trend that some have argued is probably irreversible at this point. The point here is that the 'bi-nationalism' in social and economic infrastructure is the contingent outcome of broad market forces, as opposed to clear governmental directions.[123]

The integration that is evident in this sphere is akin to the 'self-synchronisation' aimed for in the theory underlying NCW. It stands in clear contradistinction to the efforts to coordinate bi-national approaches to national security through the auspices of NORAD. In its call for a 'need to share' as opposed to a 'need to know' approach to information sharing, the BPG clearly points out the requirement to transcend the classic organisational approach to operations. As they note,

> In the past, the US Department of Defense's approach to information sharing was focused on written agreements for every type of information that might be shared and a fear that sharing would result in adverse action. In today's threat environment, actionable intelligence may be missed. The fusion of information that is required by all players domestically as well as bi-nationally is essential to success.[124]

In many ways, however, the complexity of the organisational integration required in such an approach will likely defy that vision. Many observers of NORAD are increasingly pessimistic about its long-term prospects. A bi-national team themselves, colleagues Joel Sokolsky and Joseph Jockel, Canadian and American political scientists respectively, have argued that despite the renewal of the NORAD agreement in 2006 without an expiration date, its future is 'in doubt' as each nation attempts to sort out how much of homeland defence to keep national and how much to make bi-national.[125] Former USAF NORAD staff officer, Bernard Stancati argues that NORAD is gradually atrophying given shifts in attitudes in both the US and Canada on the value of cooperation, each nation in some respects pursuing more unilateralist policies with regards to their national security.[126] Dwight Mason argues that while North American security cannot be managed alone by the US, Canada will be unable to 'sustain its participation in the long run much less take on new responsibilities' in the face of falling military resources.[127] Finally, James Fergusson notes that any bi-national venture may have to revisit the whole command structure, with Canada as the perennial junior partner: 'the symbols of sovereignty and independence, and national independence' may re-emerge, demanding such a reconsideration, the success of which he notes as being 'unlikely'.[128] Within NORAD itself, concern that the organisation was drifting away from its founding principles were evident amongst many officers. LGen. Eric Findley remarked:

> I think if you do a bi-national or tri-national information sharing, warning – that can be made to work. Is the execution piece really that critical? Not really. When you think about it, we always divide it up into Canadian NORAD Region, Alaska NORAD Region, Continental NORAD Region.[129]

Officers on the BPG noted:

> There are clearly cases where a bi-national response or bi-national programming is the optimal solution. But there are also clear cases where a bi-lateral

> response works best and is most efficient. And of course we have to recognize the sovereign right of each country to do things unilaterally, because we have different relationships with different countries in the AOR.... So you have bi-national surveillance and warning but you have bi-lateral execution.[130]

MGen. Angus Watt was even more blunt:

> (NORAD is) being challenged by evolutions in policy. In fact, the evolutions in technology should allow us to do that even better, to seamlessly integrate our two nations in the defence partnership. Unfortunately, it is the evolution in policy between our two governments that's pulling that apart. And NORAD is becoming less an integrated command and more a coalition. And when you become a coalition, then the relationship changes. NORAD is evolving towards the coalition and there is a great deal of mission competition with Northcom.[131]

Others hoped that the two countries could return to a situation where 'Canada and the US look at the threat that affect both of us, rather than defending the land border that separates our two countries'.[132] Finally, one self-described 'old time' NORAD person remarked:

> Now you have a situation where you ask 'Well, is this an American problem or is this a Canadian problem?' and before, there was no such thing. It was a NORAD problem … and how do we solve it? But now, it's much easier … since 9/11 to look and say 'well, this doesn't really involve the United States. It's a Canadian problem.' And so … the information wasn't shared, or the chain of command changed, or who had the authority to make that decision changed.[133]

And it is the growing complexity of the security environment that is challenging not the old notion of a unified continental battlespace, but rather the functional ability of commanders to coordinate a bi-national response. Commenting on Operation Noble Eagle, one observer commented:

> although unity of effort as an ideal is a common attribute among all participants in Operation Noble Eagle, achieving the coordination and synchronization necessary to realize tangible unity is practically impossible for an operation of this magnitude, duration, and importance. With thousands of federal, state, and local agencies participating in efforts to protect the US, military commanders face similar hardships to those leaders of multi-agency campaigns involving MOOTW.[134]

The collapse of NORAD has been long anticipated by many. The emergence of ICBMs, the creation of Space Command, and the rise in interest in missile defence have all historically provoked analysts on both sides of the border to question the long-term viability of the organisation. As Sokolsky and Jockel note, 'from its inception, NORAD has been just one step ahead of changes in technology and strategy that threatened to end the command's utility for the US'. Only the fact that such changes did not impede American interests and thus NORAD renewals were relatively non-controversial for it (as opposed to its Canadian partner) favoured the organisation's longevity in the face of these developments.[135] However, the US cannot remain disinterested in its own homeland security, where the threats may emerge from within its own society as much as from distant lands.

While LGen. Joseph Inge (USA), the deputy commander of Northern Command as well as the Vice Commander of the USELEMENTNORAD could remark at a conference in 2005 that 'NORAD HQ in Colorado Springs is an example of the partnership and spirit of mutual cooperation that will remain necessary for defending our homeland', RAdm. Ian Mack (CF), then Chief of the Canadian Defence Liaison Staff in Washington DC came to a more equivocal conclusion that NORAD renewal would reflect a Canada–US relationship that was 'bi-national where necessary, but not necessarily bi-national'.[136]

To return at last to the issue that motivated this examination, the NORAD experience has key lessons for CAOCs. In some respects, the technical challenges confronting CAOCs are the same as those confronting NORAD in its air defence mission. The head of the US Army Space and Missile Defense Command, LGen. Kevin Campbell remarked

> If you look at the time line of a missile launch, there really isn't time to have a pleasant discussion over who's going to do what. It has to be decided in advance.... When we look at our allies, we shouldn't be saying, 'let's include them.' We should be saying 'We can't do this without them.'[137]

Indeed, the same general has called for a 'NORAD East', although he noted that while a European version of NORAD is 'probably a good idea', there obviously would be constraints on information sharing that would have to be 'worked through'.[138] In such an understatement many a policy pitfall is obvious.

The example of NORAD shows the bankruptcy of the capital investment/multi-tiered alliance argument that often accompanies criticism of NCW. NORAD evolved an early form of NCW in the 1950s, one in which the US ensured its Canadian partner was kept abreast of technology developments.

The slow degradation in cooperation between Canada and the US has less to do with capital investment and more to do with policy coordination and strategic differences over the role each nation sees for itself in the world. The basis for the 50 years of successful cooperation between the 'elephant' and the 'mouse' was a common appreciation of the threat that confronted the continent, and a common agreement that both countries had to work closely together in order to deal with that threat. Even here, however, as close as each country was, there was always mutual suspicion at the strategic level that Canada would abandon the US in a crisis, or that the US would drag Canada into a war against its will. The fundamental lack of strategic trust between Canada and the US remains, and if anything, may have actually grown since 9/11. Future plans for NORAD actually envisage less integration of military forces, in a trend that is the reverse of what the two countries are experiencing in almost every other sector of their national existence.

Second, despite widespread recognition that government departments need greater levels of information sharing in order to address the complex post-9/11 security environment, such admonishments remain at the rhetorical level. There are sound technical reasons for this outcome. The desire for information sharing is based on an Internet experience that is not directly transferable into a governmental environment. Law enforcement agencies may be able to effect limited forms of information sharing when it suits their purposes; however, crossing institutional boundaries is often alien to these. The need to avoid compromising jurisprudential information conflicts with the broader goal of enhancing *bi-national* security. As much as military commanders and defence officials will continue to encourage such cooperation, because they must, the best that can be hoped for is not dissimilar to the unclassified information sharing that was ultimately arranged between naval commanders in the Persian Gulf. This is decidedly less than the level of cooperation that was possible between the RCAF and USAF during NORAD's heyday.

As LGen. Campbell noted,

> We still cannot pass (to British and Australian personnel) the type of information to really do the job.... So I worry that, as we bring on other partners and we try to do a combined Command and Control center, it will be somewhat haves and have nots in the command center.[139]

Here again, the primacy of politics over operations asserts itself in a critical fashion. The experience of NORAD raises real questions about the long-term success of greater levels of integration of coalition partners in CAOCs. High levels of integration in the areas of air defence and on some space matters have been possible between Canada and the United States, simply because their interests overlaid each other so directly. Over time, as the strategic environment began to shift, even high levels of operational respect between the two air forces could not protect the relationship from the strategic forces buffeting it. NORAD is likely to persist, at least in terms of its air defence role. While ITWAA is currently not threatened by the missile defence decision, as the US missile defence architecture begins to mature, more and more aspects of that will probably begin to fall outside the NORAD architecture, at which point Canada's vestigial role may become more a hindrance than desirable.

6 Information, geography, mobility, and coordination

Land operations in digital coalition battlespaces

> There is no substitute for a waterproof map you can stick in your pocket.[1]
>
> British Army Officer, Operation Iraqi Freedom

Vincent Moscoe, in *The Digital Sublime*, observes that the themes of the end of geography, history, and politics are central to much of the literature discussing the emergence of the information age. The same themes are evident in military literature: the constraints that geography has always imposed on military operations, particularly land operations, have been transcended by new means of connecting military units; our age is a revolutionary one in which the old constants of the past no longer apply; technology has transcended politics to such a degree that one need not even consider the strategic underpinnings of a conflict – modern digital forces will simply overwhelm any opposition in a 'rapid and decisive operation'. Relying simply on a dataset provided by the advances achieved in the air and at sea, it is easy to arrive at such conclusions, but the real crucible of warfare has always been on land. Here it is far from evident that geography, history, or politics have been transcended by modern military technology. Indeed, in terms of coalition operations, all three issues are critical to their success or failure.

In current operations, the predominance of American military power, together with the uncertainty introduced by globalisation, discussed in Chapter 1, combine in a pernicious way with the natural infighting discussed in Chapter 3 that occurs between military partners in both coalitions and alliances. The primacy of risk and its management have resulted in the limited approaches that all nations, aside from the United States, bring to the coalition battlespace. The inherent complexity of the land battlespace in failed states, urban environments, and ungoverned tribal areas raises the risk of operational failure. Aggressive operations result in high numbers of casualties and a high probability of fratricide, undermining the willingness of the coalition partner to continue its participation. Conversely, constrained limited operations can be easily taken advantage of by determined opposing forces, a situation ultimately resulting either in operational stasis like that experienced in the Balkans during the mid-1990s and currently in Darfur, or in a desultory withdrawal of coalition forces as occurred in Somalia and Rwanda.

The combination of this witches' brew of factors together with the shrinking military capabilities of all developed nations contributes to a growing reliance on America's global precision military capabilities, a process which is in turn reinforced by their continuing efforts to outpace both friends and competitors in this game. The end result is a growing mismatch of capabilities between the US and those it seeks assistance from. American unilateralist tendencies are enhanced, thus by this twin process consisting of the generalised reluctance to engage in support of international peace and security and, second, the increasing difficulty in operationalising military cooperation due to the rapid advance of digital technologies. Perversely, the very technologies which are meant to enhance communication and collaboration are those that may undermine wider military cooperation.

Geography and strategic power

The US faces two critical geostrategic challenges. American strategic power stems from the relative geographic isolation of North America. As the axiom goes, America has oceans to its east and west and weak neighbours to its north and south. In its weak formative years, these geographic features allowed the US to develop in relative isolation from the strategic concerns that have historically confronted small powers. The same geographic isolation that permitted the development of the US as a continental power, today, complicates the global exercise of its own land power. In the first half of the twentieth century, engaging the US in military activities was a complex process of convincing an isolation-minded domestic audience of the need for engagement and then getting the forces to where they were needed. During the Cold War period and after, the first issue could be taken as a given, but the trouble of moving US forces in sufficient numbers to where they are needed continues to be a most difficult strategic challenge. In its future strategy, IT plays a key role in reducing this problem, raising the 'fungibility' of American land power. Yet even if the promise of technology is realised in easing its capacity to project land power abroad, the same developments may compromise its ability to work with others. The second conundrum, then, for the US is that if its transformation plans are successful, they will compromise its approach to working in cooperation with like-minded groups, as it has preferred to do so throughout its history. If America fails to realise the promise of IT in the land battlespace, while its coalition approach may benefit, its quest for global strategic dominance, with its project to underwrite international security, will be greatly compromised.

As America's *Transformation Planning Guidance* notes:

> Transformation is 'a process that shapes the changing nature of military competition and cooperation through new combinations of concepts, capabilities, people and organizations that exploit our nation's advantages and protect against our asymmetric vulnerabilities to sustain our strategic position, which helps underpin peace and stability in the world.'[2]

In 2001, the *Quadrennial Defense Review* laid out six objectives for America's military. They were:

1 Protecting critical bases of operations (U.S. homeland, forces abroad, allies and friends) and defeating CBRNE weapons and means of delivery will ensure our ability to generate forces in a timely manner without being deterred by adversary escalation options.
2 Projecting and sustaining U.S. forces in distant anti-access or area-denial environments and defeating anti-access threats will enable us to preserve and utilize the most effective avenues of approach while rapidly engaging adversary forces.
3 Denying enemies sanctuary through persistent surveillance, tracking and rapid engagement with high-volume precision strikes will permit the United States to prosecute a rapid campaign that reinforces deterrence by denying any adversary hope of achieving even limited objectives, preserving escalation options or maintaining command and control of forces over an extended period.
4 Assuring information systems in the face of attack and conducting effective and discriminate offensive information operations will deny the adversary hope of exploiting a new dimension of the battlespace as a low-cost and powerful asymmetric option while providing us an unwarned strike capability that contributes to a broad, simultaneous and overwhelming range of effects that increases the likelihood of rapid collapse of an adversary's will to fight.
5 Enhancing the capability and survivability of space systems and supporting infrastructure will provide sustained, protected, global C4ISR capabilities that permit rapid engagement of American power and reinforce deterrence by promoting earlier warning of adversary intentions while denying the adversary similar capability.
6 Leveraging information technology and innovative concepts to develop an interoperable, joint C4ISR architecture and capability that includes a tailorable joint operational picture will guarantee our combat leaders decision superiority and enable our forces to maneuver effectively to gain positional advantage, avoid battlefield obstacles and successfully attack the adversary even in the face of numerically superior forces.[3]

Taken together, the two documents lay out the underpinning for American defence strategy at the turn of the century. The concepts laid out in each travel along two essential themes. First is clearly the end of the Cold War and the sudden emergence of America into a position of strategic primacy with the collapse of the Soviet Union. The second theme is the emergence of information as a fundamental aspect of modern society. We can see these two themes at work in the 2001 QDR's six objectives: one through three speak to the challenges and opportunities of being the sole superpower. While it commands the commons, America is geographically distant from unstable regions. The last three objectives speak to the informational advantages that America hopes to minimise that geostrategic issue.

The challenges identified by the 2001 QDR affect the Army the most. Information and geography underpin the effort of America's army to modernise itself in this new era. Alone of the three services, the Army has faced the most significant challenges in dealing with the role of information in the battlespace. The Army remains irreducibly a human centric organisation, demanding huge numbers of both personnel and equipment. On top of the enormous shifts in technology, the Army is also dealing with the tremendous operational challenges of the War on Terror. President Bush has described the challenge of transformation as being 'like overhauling an engine while driving 80 miles per hour. Yet we have no choice.'[4] The absence of choice that Bush refers to, points not only to the relentless advance of technology, but also the fear that America's privileged strategic position is subject to eventual erosion without continuous action. Second, while the Navy and Air Force operate relatively unchallenged in their specific environments, American land power is, by its very nature, contested: there is no commons to command. The land is not politically neutered as international waters and airspace can be. Finally, both the Navy and Air Force are by their very nature 'expeditionary' services; their firepower is inherently mobile while the Army faces the challenge of moving its forces to areas of strategic concern.

Bridging America's geostrategic conundrum: information, armour, and mobility

The beginning of the 1990s seemed to promise a new era of military efficacy in the example of Desert Storm. At the close of the Vietnam war, the Army sought to return to its roots in conventional conflict. An intense theoretical exercise in analysing and applying the lessons of large-scale conventional war from the Second World War to the problems of fighting the Soviets on the Central European front resulted ultimately in the development of AirLand Battle. Desert Storm realised 15 years of doctrinal and technological modernisation in the application of this theoretical process, suggesting, to many, opportunities similar to those of the introduction of mechanised forces at the close of the First World War. As in 1940, warfare seemed to have entered a new practical realm in the art of the possible.[5]

The 1990s' visions of the future of military power came quite grounded in the mythologies of the past, especially the fear of missed opportunities; May 1940 underlay much of the *Sturm und Drang* that was to erupt in the debates over RMAs and ultimately, transformation plans. Just as Britain and France had failed to understand properly the technological implications of armoured warfare, and so had been decisively defeated by Germany in May 1940, there were fears that information technology was leading to an evolution away from the historical industrial model, whose achievements would be rendered irrelevant.[6] Indeed, in the period of the mid- to the end of the 1990s, several academic works appeared re-examining the interwar period as a source of military innovation, highlighting the successes and missed opportunities that presented themselves during this formative period.

Throughout the 1990s, the Main Battle Tank, symbolised ideally by the iconic M-1A1, assumed the role of representative dinosaur: powerful, but heavy and difficult to deploy. The dilemma facing the Army was summed up with the observation that American 'heavy forces have limited strategic deployability and our light forces have limited tactical utility'.[7] The infamous and ill-fated deployment of Apache helicopters to Albania in the midst of the Kosovo conflict, symbolised all that seemed wrong about the strategic utility of the powerful but heavy US Army.[8] The hope was (and still is) that 'transformation will take care of that disconnect'.

In resolving the armour/mobility geostrategic dilemma, the Army seeks to take advantage of the opportunity that new developments in information technology promise. Effectively, enhanced situational awareness, courtesy of networked sensors and decision makers, would allow the Army to trade its armour for greater manoeuvrability in strategic, operational, and tactical terms. As part of the Full Spectrum Operations concept, the Army would be able to undertake 'Dominant Manoeuvre'. Superior knowledge of the battlespace would enable land forces to achieve dramatic impacts in terms of their ability to both position themselves and engage enemy forces. Multidimensional assaults from land, sea, and air taking place simultaneously throughout the entire volume of the battlespace would not be easily countered by opposing forces, creating paralysing effects in the minds of enemy decision makers.[9] The growing lethality of long-range precision weapons would continue the expansion of the battlespace by enabling the dispersion of scarce military resources over larger and larger areas throughout it. Precise knowledge of the friendly and enemy forces would permit units to customise themselves to specific operational demands and reduce the need to 'fill space with forces and direct fire weapons'. As FM-1 put it, 'The goal of future Army operations will be to simultaneously attack critical targets throughout the area of operations by rapid manoeuvre and precise fires to break the adversaries will and compel him to surrender.'[10] A so-called 'Quality of Firsts' would be generated by these relationships: information-age land forces would be able to 'see first, understand first, act first, and finish decisively'. For the first time in the history of warfare, land forces would be able to pick and choose the time and place of engagement.[11]

However, such information-age land forces would be qualitatively different from those of the past. The information systems that would make them powerful offensive forces would come at the expense of armour and to a certain extent, self-reliance: these units will be profoundly reliant on the coordination of services from other, typically more distant supporting units. While information would alleviate the dichotomies raised between light and heavy forces, it would not eliminate them altogether. Lacking sufficient armour to defend themselves independently, the new units would need to draw on the capabilities of other forces, possibly in different services, to provide complete force protection. Information distributed through networks would heighten the offensive power of land forces, but would in turn become absolutely critical for their defence as well in the key role of coordinating the efforts of dispersed and individually weak forces.[12]

In order to realise its vision for future warfare, the Army has pursued a number of programmes, most visibly in terms of the Future Combat System (FCS), Stryker Brigades, and its Landwarrior infantry system. Both the FCS and Stryker Brigades emerged from the reforms General Eric Shinseki, concerned by the implications of the failed Task Force Hawk mission, sought to implement. Each forms a part of the other in terms of achieving the objective of a more manoeuvrable army. The problem of wholesale modernisation in armies relates to the challenge of shifting the technological base away from the older model to its replacement. The shift from so-called 'legacy' forces to future systems is more troublesome in many ways than in other services given the larger number of intensely interrelated systems that must be modernised. As such, the Army has proposed a series of steps through which it will pursue its shift from industrial-age based forces to the goal of an information-age Army.

Stryker Brigades were originally designated as 'Interim Brigade Combat Teams' to reflect this stepped approach. Based around a common light armoured chassis from the MOWAG Piranha family of vehicles, the Stryker Brigades rely heavily on distributed sensors to provide superior situational awareness, compensating for their relative lack of armour.[13] Although lacking in armour, when fully realised, it is hoped that the Brigade will bring the same capability as a scaled-down division.[14]

The FCS has emerged from the so-called 'Objective Force' of Shinseki's proposal. It consists of a system of eight different vehicles, also based around a common chassis and fuel efficient engine. Supporting this family of vehicles is a series of UAVs, unmanned ground vehicles, sensors, and intelligent munitions. All of these systems are to be networked together enabling not only comprehensive situational awareness of the battlespace for planning and conducting operations, but also for logistics, maintenance, and even personnel management.[15]

Landwarrior seeks to accomplish the same goals of situational awareness for individual troops manoeuvring on the ground itself. It is composed of a helmet-mounted display system integrated with a personal radio, capable of transmitting and receiving voice and data, and a optical digital sight mounted on the soldier's weapon. The entire system permits soldiers to see where their squad mates are on a digital map, send and receive formatted text messages, and use the sight to improve their marksmanship through its zoom and night vision capabilities. The integrated helmet display also allows soldiers to look and fire around corners.[16]

The experience of Iraq

If Desert Storm seemed to indicate a break with past military history, the 'major combat operations' in the spring of 2003 seemed to indicate that the character of military operations had in fact experienced a fundamental shift during Desert Storm and that the policy and capital plans pursued by the US since then were now paying the anticipated dividends.

In 1991, only artillery units made regular use of GPS, whereas 90 per cent of units were so equipped in 2003. For all its advance, Desert Storm was a sequential operation of air and land campaigns, whereas operations in 2003 were conducted simultaneously. Most impressively, a much smaller force covered a significantly larger area: 437,000 square miles as opposed to 40,000 square miles. Joint fire control improved dramatically, taking only 45 seconds on average to clear fire requests as opposed to seven hours. Last, the number of incidents of fratricide declined radically with no ground to ground incidents occurring.[17]

New command and control technology permitted high speed, dispersed operations by answering some classic tactical questions: where am I, where are friendly and enemy elements, and what are their respective statuses? This information was collected and distributed on the Army Battle Command System (ABCS). ABCS is made up of three related elements, the Global Command and Control System Army (GCCS-A), which collects and disseminates strategic and theatre information to commanders, the Army Tactical Command and Control System (ATCCS), which links the various 'functional' areas of the Army (Artillery, Logistics, Intelligence, and Air Defence Coordination) enabling plans to be developed, shared, and executed collaboratively amongst dispersed units. Last, is Force XXI Battle Command Brigade and Below (FBCB2), a GPS-enabled technology that tracks unit positions in real time and collates them on digitised maps. Effectively, in the ABCS system, ATCCS links the tactical picture of FBCB2 with the theatre/strategic picture on GCCS-A enabling operations to be planned and executed at a variety of command levels throughout the entire battlespace.[18] These systems operationalise the armour/mobility/information theory underlying the digital army.

By sharing ubiquitous information at all three levels of war, Army decision makers escape the tyranny imposed by 'battlefield geometry': geographic phase lines and time tables coordinating large-scale indirect fire missions and movements of land forces across the battlespace, which themselves were earlier manifestations of armies attempting to grapple with the increasing complexity of the battlespace and manage information in an inherently chaotic and dangerous environment.[19] In the past, these older and less dynamic information management techniques permitted indirect artillery fires and airborne close air support to be coordinated with rapidly moving ground forces, as well as reducing the potential for fratricide between adjacent ground units, long range artillery, and overhead air support. Emerging in the First World War, these procedures restored a level of mobility through coordinating the actions of multiple specialised services with advancing land forces, including critical logistical support. However, they were time consuming to organise and difficult to amend for a dynamic operational situation. Digital command and control systems break the dead hand of procedure through instantaneous information sharing, permitting on the fly adjustment to plans and operations, and rapid response by distant support units to emerging tactical situations.

The issue of the management of 'joint fires' is one that has vexed military forces, especially as the reach of various weapon systems and the operational tempo of ground forces has grown. As these two variables have increased, there has been an increasing need for joint coordination, particularly between air and ground forces, and the requisite ability to share information so as to avoid fratricide. The specific issue revolves around who controls the assignment of targets in a particular geographic space. Generally speaking, long-range target assignment is controlled by the Air Force and nearby targets are controlled by ground commanders. Where the geographic edge between those two areas lies is a matter of considerable concern that brings in a whole range of technological and administrative aspects, as well as those of a more bureaucratic nature. Although the edge has been known by many terms in the past, it is currently known as the Fire Control Support Line (FCSL).

As land systems, such as MLRS and Apache helicopters increase the reach that Army units can target beyond the traditional ranges defined by long-range tube artillery, the exact position of the FCSL is a matter of constant debate between the two services. Further, as Army units have increased the velocity of their advance, considerable flexibility has had to be built into the movement of the line by operational planners. During Iraqi Freedom, US Army units threatened to overrun the FCSL on a number of occasions. The consequences of doing so were potential fratricidal incidents from overhead Air Force assets in a fast moving and confusing ground battlespace. Finally, the target assignment process on *either* side of the line is also subject to considerable flexibility in order to deal with 'time sensitive targets'. In this environment, electronic 'killboxes' were established that could be opened and closed according to the needs of commanders on either side of the FCSL.[20]

The effect of this information sharing environment on operations is a dramatic increase of their velocity, creating novel effects on enemy behaviour. Iraqi forces were denied the opportunity to laager their forces in defended lines by the swift advance of American armoured forces. The arrival of American forces on Iraqi lines forced them to move to forestall encirclement. This, in turn, exposed them to air attack; in the words of Cordesman 'jointness took on a new practical meaning' in these types of operations.[21] This type of operational dilemma presented to Iraqi commanders was, in fact, a classic expression of manoeuvre theory. The second impact was the scale of the area of operations for all units. Third Infantry Division Commander controlled operations over an area of 200–250 km whereas the same unit operated on a frontage of only 30 km during Desert Storm. Companies were often as far as 70 km away from their controlling Brigade headquarters.[22]

Both speed and dispersion naturally demand high amounts of information, but the effectiveness of the ABCS system in Iraq provided sufficient support to permit such 'untethered' operations. Early studies suggested, rather than information overload, command and control systems enabled a 'commander centred decision process' as commanders and their staff could focus less on the collection of information and more on its analysis and synthesis.[23] Discussing the inevitable risks of this style of operation, the Land Component Commander, Lt. General McKiernan observed:

> This ground campaign to date has reflected itself in high tempo continuous operations, decisive maneuver, extended logistical support, where I accepted

> some risk in the length of our lines of communication and our logistical reach ... we have overcome that risk, and an execution of a plan that had several options in but always remained focussed on the enemy.[24]

FBCB2 played a dominant role in shaping the speed and dispersion of OIF. The system had been proposed in 1995 and later used operationally for the first time by US forces participating in stability operations in the Balkans. Initially, the system was based on high frequency radio, however, the difficulty of maintaining radio line of sight contact in Kosovo's mountainous terrain spawned a satellite-linked version of the technology that came to be known as Blue Force Tracker (BFT). The two systems initially were not connected to each other (BFT-equipped units did not see the data from FBCB2-equipped units and vice versa), although that has since been rectified. In general, the two systems are synonymous.[25]

In concept, FBCB2 is simple, but incredibly complicated in practice. Effectively, the system automates the reporting process units traditionally engaged in during combat. This has typically been a time consuming manual affair that requires units to check in regularly with status reports. A degree of error was inevitable as units became disoriented; information might be recorded inaccurately or irregularly. As operational tempo increased in headquarter units, information might not be prioritised correctly and thus fail to be recorded at all. Paper maps complicated information management, as distant units crossed on to new map sheets before others did. Radio difficulties might prevent units from reporting at all.[26] The hierarchical structure of army units imposed added difficulties as information moved up from the front, but not necessarily back down or laterally. Analogue radio networks themselves added to this rigid structure in their inherent inflexibility.[27] The entire process created as much fog and friction as it was supposed to eliminate, yet given the technology of the day, there was little alternative.

There is little to wonder, then, why FBCB2 was so welcomed by land forces, although unfamiliarity with the technology initially hindered some acceptance of it.[28] For all equipped friendly forces, the 'recognised common picture' of FBCB2 keeps track of their location, available resources, current status, any control measures they may be subject to, and their planned actions. These are stored in a database that is linked to a digitised map with clickable moving icons that display a unit's status on demand. As units move, a GPS transmitter updates the map location for the unit, while any changes to orders, resource status, and the like can be sent via preformatted messages by e-mail or chat. Recording enemy forces was still a manual process, dependent on reports from units in contact and so red icons did not move in the same manner that blue forces did. Finally, logistic units were also tracked on the system (location and status only).[29]

FBCB2 clearly enhanced the speed of operations by automating a complicated and error prone system. The stop and report system that moved information ponderously up from platoons to companies, battalions, brigades, and so on was replaced by digitised systems that required only changes in situation status to be reported while keeping accurate track of geographic locations. Most importantly, information previously limited to printed reports was now stored digitally and available to all. As such FBCB2 vastly improved situational awareness of all so-equipped units. With those units operating the satellite-enabled BFT, units could keep track of theatre-wide movements of forces, permitting them to see whether they were keeping pace or falling behind operational advances, even when voice communications had been lost. This is a significant advance over Operation Desert Storm where tactical information sharing between units significantly impeded the movement of VII Corps.[30] Further, information was no longer limited to vertical chains of command; units could now examine the status of those located on their flanks as well, thus enabling greater coordination laterally.[31]

As it did at sea, chat played an important role in sharing time sensitive information and maintaining situational awareness. Again, in many ways, chat provided an organisational work-around that permitted 'a quick flexible way to share information that ignores traditional hierarchy and architecture'.[32] Satellite-enabled BFT terminals allowed rapidly advancing units to remain in contact through rudimentary e-mail and chat features, even as they moved beyond the range of their VHF radios. Such functionality created 'oneness' within coalition forces. The ubiquity of information generated the confidence to allow units to move farther and faster than in previous military campaigns.[33] As the commander of the 4th Infantry Division, the only fully digitised formation to fight in Iraq, noted:

> Our ability (to) track every tank, every Bradley, every howitzer, and we know exactly where they are, wherever they go. I have that capability to do that from my operations center. So it makes it much easier for us to understand where our soldiers are, how they coordinate their cordons, how they coordinate their final attacks during these raids.[34]

Some have compared the functionality of FBCB2 with that of the Internet, noting its self-adaptive character and ease with which information can be shared over it. As units logged into the system, it updated the database, enriching the operational picture. The technology is spreading rapidly throughout the US Army, 210 terminals were employed in Afghanistan, 1,242 in Iraq – including US Army, USMC, and UK units.[35]

Conceptual problems: geography, politics, and information in the land battlespace

Despite the obvious success of FBCB2 during conventional operations in Iraq, there has been growing concern that the network vision underlying the information-mediated armour/mobility compromises on which Army modernisation is predicated is affected by service specific assumptions about the character of the battlespace environment. Predictions of information superiority, leading to dominant battlespace knowledge, may in fact be reliant on a model that springs largely from the nature of air and sea warfare.[36] The operational environment differs significantly between land forces and those using air and sea spaces. The latter are *relatively* transparent mediums that are generally devoid of targets. While naval combat is certainly affected by topography to a degree when considering the complexity of anti-submarine warfare and inshore littoral operations, both the number of contacts a naval commander must track and identify as well as the number of sensors used to accomplish this pales in comparison with that of his land counterpart. Further, the nature of a contact is far easier to characterise in both the air and at sea given their limited numbers as well as the general absence of terrain in which to hide. Air units seek to conceal themselves in the vastness of the airspace itself. Naval units behave similarly; land impacts on them only in the littoral areas where they begin to be confronted with problems familiar to land commanders. In the littoral, however, the nature of the tactical problem emerges from the complex nature of the land itself, rather than the nature of the sea.

Land operations differ significantly from those conducted in the air and at sea. Principally, the difference is found in the number of moving parts involved in land operations. While an airstrike may involve the coordination of many different types of units over a wide geographical area (AWACs, tankers, ground-based air defence, combat air patrols, aircraft flying suppression of enemy air defence, electronic warfare, offensive counter air, as well as the actual strike mission to list just a few), the number of participating units number in the hundreds at most; at sea, the number is even smaller, typically involving tens of units. On land, thousands of units and individuals are involved in even the simplest and least dangerous of operations. Worse, units and individuals operate in an 'austere' environment where they must bring along their own logistics, maintenance, communications, and medical support with them, each of these moving about the battlespace and complicating the overall operational picture. Air bases are fixed locations typically well supplied with all these facilities and ships at sea can put into port or are large platforms on which all these functions may be easily based.[37]

Leaving aside the inherent organisational complexity of coordinating the thousands of moving units and individuals to achieve a common objective in the face of danger, which FBCB2 seems to have resolved to a considerable degree, an opponent's behaviour and his use of the battlespace are conceptually difficult problems to digitise. The land battlespace varies in terms of its 'density'; urban or jungle environments are dense in their nature in ways that desert environments are not. Visibility may range over a matter of kilometres in deserts, conversely, in cities and jungles, visibility may be only metres. In dense environments, a land force must invest a significant amount of energy in the form of reconnaissance in order to distinguish the enemy from the surrounding environment. The challenge in this task is a combination between two key aspects. First are geographical features (not necessarily strictly limited to physical terrain) which can be exploited by opponents through camouflage or other deceptive measures in order to conceal themselves. Second is the behavioural nature of the enemy.[38] Gross military capability is made up of a variety of factors including equipment, doctrine, as well as terrain. How an opponent takes advantage of the opportunities those factors present is very much dependent on issues of strategy, leadership, and morale.

Some advances in tracking red forces on digital maps have been made. FBCB2 clearly aided coalition forces with theatre-wide situational awareness, however, its impact on resolving these red force problems was more limited. Commanders complained that red force data on the system were unreliable as they had to be entered manually and did not move about the digital map because they were not automatically updated as blue force information was. JSTARS uses its synthetic aperture radar to generate 'moving target indicators' (MTI) from ground-based targets. However, MTI data contained no information on the nature of the target – is it a friend, foe, or a neutral. This is an assessment that goes beyond the physical geographic parameters of the opponent; his location in time and space in other words. Since 2003, linkages have been established between FBCB2 databases and those created and maintained by JSTARS aircraft. Interoperability between these two systems means that MTI and FBCB2 data can be merged so that red force data entered into the FBCB2 system can be tagged with existing MTI data and vice versa. This enables red force data to move on the FBCB2 system in the same way that blue force data does.[39] The enhanced situational awareness that FBCB2 systems promise enable conventional militaries to operate in urban areas with greater degrees of confidence by reducing the impact of the enemy's superior knowledge of the terrain.[40]

The physical features of target differentiation in even straightforward conventional engagements are also considerably more complex than at sea or in the air.

> The ability to characterize armour versus other military systems seems to have remained a problem (in OIF), as did the ability to find well dispersed systems like aircraft and individual surface to air or surface to surface missiles that were not actively moving or emitting.[41]

Human eyes are the principal sensors on the land battlefield; merging the data generated by them with digital systems is especially challenging. The number of contacts in a land scenario confronts track managers in keeping the data consistent. Differences in map readings, latency in tactical intelligence reports, movement of units and vehicles, all conspire to generate significant uncertainty in the nature of the data. All of these challenges are also present at sea and in the air, however, it is the number of sensors and contacts to be managed in the land scenario as distinct from the other two environments that complicates the process for land situational awareness networks.

However, geographical topography is only one aspect that shapes the overall knowledge of the land battlespace. Insurgents, terrorists, and even conventional forces can seek to enhance battlespace density by blending in with civilian populations and exploiting an area's political geography, complicating the technical process of situational awareness.

Further, enemy behaviour and intent form the other crucial half of battlespace density, an assessment in complex battlespaces that extends beyond the simple categories of friend/foe/neutral. As such, analysing the density of any given battlespace inevitably requires judgements that are necessarily political and subjective rather than objective in their basic nature.

Enemy behaviour is shaped by a whole series of subjective and thus inherently uncertain factors at the strategic, operational, and tactical levels of warfare. Aspects such as political will, the nature of sought after objectives (and the inherent wisdom of such goals), how such objectives are operationalised in military plans, down to specific tactics, techniques, and procedures will all inform how an enemy behaves on the battlefield.[42] None of these aspects are amenable to a red force tracking icon. In fact, some commanders have complained of the growing cult of 'iconology' or 'blobology' that stems from the influence of FBCB2. An icon on the FBCB2 screen is simply a symbolic representation of a particular enemy unit. However, it becomes suffused with operational meaning that may be inappropriate or inaccurate in many ways, assuming particular models of behaviour, organisation, and technology. In Iraq

> time and again, large conventional formations would crumble in the face of American assault while small bands of Iraqi irregulars offered intensely fierce resistance. In this context, an icon was essentially meaningless because it told a commander little about what type of enemy contact he could expect or what that enemy's intention was.[43]

By its very nature, the battlefield is subject to fundamental uncertainty.

To a certain degree, the speed and dispersion that new situational awareness technologies contribute to modern operations are also a source of this uncertainty. While they contribute to the 'Quality of Firsts' championed by the US Army, they do so only in ideal environments. Operational tempo may compromise the ability to see and understand first. One study points out that a quality of firsts inspired operation assumes that 'a tactical commander has the flexibility to modulate his unit's tempo and manoeuvre his formation after he has a good understanding of the enemy'. In operations that demand a high operational tempo, the opposite may be true; commanders must act before all the information is in. The history of land operations is one where commanders must fight for battlespace information, rather than having it before engagements.[44] Despite having revolutionary situational awareness technology, land commanders may face even greater uncertainty than previously because of the operational speed and dispersion that has been enabled by the same technology.

Dispersion calls for a high degree of situational awareness between friendly forces in order that units can both provide each other stand off support and prevent the enemy from exploiting the physical gaps that will exist between them. Decentralised enemies exploiting the physical as well as the political geography of the battlespace present considerable challenges in this regard. In order to counter the high speed operational tempo that networked conventional forces bring to the battlefield, encapsulating sensor, shooter, and decision maker into individualised and networked units ensures that insurgents and terrorists can always operate within the Boyd OODA loops of their hierarchical conventional opponents. 'A return to a warrior ethic may be the solution to countering a network enabled force', concludes one British military officer.[45] This outcome poses distinct challenges for lightly armoured and highly dispersed forces. Outnumbered and outmanoeuvred by nimbler light forces, they will be reliant on overwhelming fire power, much of it indirect, in order to counter them. While dispersed digitised forces may be able to survive in such hostile environments, the resultant damage of such reactions may ultimately undermine the wider operational and strategic goals sought after, a key aspect of the insurgent's tactics.

The fundamental nature of land warfare is shaped by the complexity of battlespace density. This complexity is resistant to approaches that seek to reduce the combat problem to a digital physical geography of location, available resources, and operational status. One American officer has pointed out the inherent contradictions that underlie the assumption of an uncertain, dynamic, and complex strategic environment and a tactical environment shaped by information superiority.[46] There is a direct relationship between tactical behaviour, operational plans, and strategic objectives. Data that easily fit into digitised data models poorly capture the relationship between these aspects. Such problems are not resistant to human analysis, however the means for grappling with such issues typically rely on a wide variety of information sources, 'unstructured data' that are not easily shared between formal digital systems (but which may be easily communicated between human operators, provided they are able to contact each other). Recent multinational experiments examining issues of interoperability between Western military forces have been struggling to come to grips with this essential problem.[47]

John D. Nelson suggests that the Army's new technology and the warfare theory that underlies it may be extremely effective in 'conducting decisive operations', but of less use in 'securing the peace'.[48] The observation reveals a fundamental problem related to the essential political foundations at the root of land warfare. America's information-age land warfare model of decisive battlespace knowledge is effectively based on a materialist understanding of warfare: the adversary is overpowered; he is unable to achieve his objectives; he capitulates or is destroyed. While the theory draws its inspiration from manoeuvre warfare theory in its discussion of the creation of paralysing psychological effects that ultimately lead to defeat, the implicit model of warfare is still solidly locked into an attritionist mindset, as revealed by this materialist focus. As Clausewitz points out, the political objectives over which war is fought are themselves not part of war. It is these political foundations of war which sustain opposing forces even in the face of total annihilation, and typically also form the basis for intra-alliance and coalition disputes.

British efforts to network military forces

> You are running so fast that we might not be able to keep pace, or even catch up when you have attained your goals.[49]

As America's principal military partner, the UK has a strong interest in keeping pace with American military developments. With the third largest military budget in the world, according to the Swedish think tank, SIPRI,[50] the British worry about not being able to keep pace with the United States clearly puts the problem of coalition interoperability on the battlefield into stark context. If the British cannot keep pace, what hope is there for the rest of America's military partners?

Like many militaries from developed nations, Britain is vigorously pursuing a variety of network projects under the aegis of its 'Network Enabled Capability' (NEC) vision. NEC first appeared in 2002, four years after NCW had become part of the public debate, which may indicate a certain ambivalence on the part of the British military to follow blindly down the path the US military was blazing. Indeed, the adoption of the term Network *Enabled Capability* as opposed to Network *Centric Warfare* indicates a level of ambivalence towards the more expansive concept promoted by the US. However, looking at the language of the document, it is difficult to detect this. For example, 'NEC offers decisive advantage, through the timely provision and exploitation of information and intelligence to enable effective decision making and agile actions'. It argues for creating a system that can exploit the latent power that exists between the seams of all three British services, thus enabling effective joint fires and an ability to engage time sensitive targets. Indeed, expropriating many of the terms from NCW, it argues that 'NEC will enable Decision Superiority through Shared Situational Awareness within task-oriented communities of interest that exploit collaborative processes in a single Information Domain'.[51] Some have argued that NEC rejects NCW on the basis that the self-synchronisation aimed for by the latter 'is a step too far and not in accord with current doctrine'.[52] However, it is not especially clear in even this analysis, how mission command fundamentally differs from concepts of self-synchronisation.[53] At best, the differences between NEC and NCW remain poorly articulated, again leading to the speculation by some that the difference is solely a 'typical British reserve' for a 'typical American enthusiasm'.[54]

The UK is pursuing a variety of network projects including the Defense Information Infrastructure (DII), which is a SIPRNET equivalent within the British Ministry of Defence, the Joint Operations Command System (JOCS), which operates very much like the Global Command and Control System, Skynet 5, a military communication satellite, the Future Rapid Effect System (FRES), which replicates some of the capabilities of the FCS, and the tactical Bowman radio system.

Like the US Army, British land forces are also based around a heavy armoured capability developed for major land combat on the European continent rather than for a rapid global expeditionary contingency. In this, the UK suffers from many of the same geostrategic challenges in the use of land power to support international order as does the US. FRES is meant to deal with some of these issues, enabling UK land forces to move rapidly in order to 'nip (a) crisis in the bud'. Like the FCS, FRES initially was intended to fit within the C-130 flight envelope. FRES differs from FCS in that the UK does not envision replacing its entire force with FRES, only developing a medium weight capability that might operate alongside heavy forces in flank and rear area operations, as well as operating in 'complex terrain'. FRES would also have some independent capability for 'intense OOTW' and 'achieving rapid effects short of warfighting' in crisis situations.[55] In this, FRES is envisioned as a supporting capability for heavy UK forces, which will themselves be replaced by follow on heavy forces in the form of the 'Future Ground Manoeuvrability Capability' in 2035.[56] However, like FCS, the weight limit for FRES has been steadily creeping upward as information has failed to be able to compensate for a lack of armour. Commenting on the experience of Iraq, BGen. Lamont Kirkland noted recently 'Whereas before a certain degree of risk was acceptable – and expected to be compensated for by operational procedures designed to reduce vehicle exposure to expected threat – that degree of risk has now been significantly lowered', requiring a greater amount of armour and thus vehicle weight.[57] As Lord Drayson noted in the House of Lords on 23 April 2006, 'Force protection in theatre now has a higher priority than strategic deployability.'[58]

The problem-plagued Bowman radio system replaces the analogue 'Clansman' radio system in service since the 1970s. Incorporating automated digital encryption and GPS positional location, it is hoped that Bowman will both increase operational tempo as well as reduce fratricide. With its ability to report both position and status of units, Bowman replicates many of the functions of FBCB2.[59] Positions of ground forces will be reported in real time, it will support e-mail, and permit limited web browsing as well.[60]

Bowman incorporates a similar programme to the ABCS, the Combat Infrastructure and Platform Battlefield Information Systems Application (CIP), which is itself divided into two components. First, the Combat Infrastructure is composed of the Common Battlefield Application Toolset (ComBAT) and the Digitisation Battlespace Land Infrastructure (DBLI). ComBAT essentially replicates the functionality of the American Army Tactical Command and Control System (ATCCS) in linking the functional areas of the Army together for collaborative planning and battlefield management. DBLI provides the hardware and software backbone. Second, the Platform Battlefield Information System (P-BISA) integrates ComBAT and DBLI into armoured fighting vehicles like the Challenger tank and Warrior fighting vehicle.[61]

The Bowman project is understandably complex. It will be implemented in seven brigades and one Commando brigade, 15,700 vehicles from Challenger tanks to Land Rover jeeps, but it will also be installed in naval vessels, Chinook and Merlin helicopters, and links to Apache helicopters as well.

At the beginning of OIF, British strategic/operational communications architecture was at an early stage of development. The Operational Strategic Communications ARchitecture (OSCAR) provided network communication between strategic headquarters in the UK and fixed and secure sites deployed abroad. This was supplemented with a rapidly developed point to point network of Iridium satellite phones. Despite the importance of satellite communications for network operations, the UK experienced a significant shortage of satcom channels which necessitated 'command involvement' to secure the required assets, only confirmed nine days in advance of the commencement of operations.[62]

JOCS was deployed on OSCAR. JOCS functions in a similar manner as SIPRNET in that it provides a data link and messaging capability between all UK headquarters, including those deployed abroad, albeit with more limited situational awareness at the time. In addition to JOCS, the UK also engaged CENTRIXS-X (Aus-UK-US) and was granted limited access to the SIPRNET through the intermediary of American Foreign Disclosure Officers.[63]

The close access the UK obviously enjoyed from its SIPRNET permissions was further reinforced by strong liaison connections within the CENTCOM staff itself. Air Vice Marshal Glen Torpy and Rear Admiral David Snelson served in senior command positions within CENTCOM, the latter as the Deputy Coalition Maritime Component Commander. Because of the relative size of the land campaign, UK land commanders did not serve in such senior positions within CENTCOM, but were delegated under the command of the 1 MEF.[64]

While the British land commanders may not have had the same level of authority within the senior CENTCOM staff, their position within 1 MEF was critical for the conduct of operations in the Al Faw peninsula and the advance towards and occupation of Basra as the US Marines accompanied British forces in their own advance inland. What made this relationship particularly critical was the long training relationship the British Army has had with American Marines giving each force a high degree of professional familiarity with the other. Furthermore, the British Army was able to take advantage of American long-range fire support through the assignment of Marine 'Anglico' air to ground coordination teams and the 'brigading' of British artillery with Marine batteries.[65]

British land units were also supplied with a limited number of FBCB2 (in this case, not the satellite BFT versions). However, only 45 units were made available, limiting the downward distribution of sets to battlegroup headquarter levels only. In this case, British units were not operating across the same ranges and high operating tempo as American units were; existing VHF communications between British land forces was sufficient for command and control.[66]

'Patchwork enabled capability'

Just as the NEC document post-dated the appearance of NCW by four years, the readiness of the British military to conduct networked operations was debatable at the start of OIF. The Bowman radio system had been experiencing significant problems in definition and design that meant it was not ready to be deployed until 2005 (and then, only under a 'spiral' process where more and more capability was added in successive deployments),[67] and OSCAR was a last minute 'urgent operational requirement' procured given that the 'Cormorant' system was still in development. According to one assessment, operational command and control was 'still based on handsets and conference calls'.[68] The speed with which OSCAR was acquired and the relative lack of training before it had to be employed in theatre lead to several systems failures, including one incident which appeared to commanders to be a cyber attack at a critical moment, although was later revealed to be an unauthorised equipment alteration by a contractor in London.[69]

Lacking the technical capability meant that the British Army entered the Iraq war with a highly 'stovepiped' command and control architecture, unlike that which had evolved within the US military over the previous decade. Whereas American army formations shared information amongst themselves horizontally as well as vertically, there was little communication laterally across similar British organisations, a feature that became more and more pronounced as one worked downwards in the organisation of that army. Thus battlegroups in different brigades did not talk to each other and neither did companies in different battlegroups. While good local situational awareness was possible under such conditions, it meant that a wider picture of the battlespace was absent in many cases. The installation of FBCB2 rectified some of this problem, but only for those units so equipped, typically headquarters given the few units available to British forces.[70] At the company level, there was no situational awareness beyond the immediate area of operations.[71]

A lack of sufficient air resources to provide completely for dedicated British air cover highlights the peril non- or semi-networked forces confront in digitised battlespaces. Rare for a typical coalition partner, the British deployed with some organic air capability, but not enough to meet British requirements on a continuous basis. The need to rely on American assets for close air support pointedly illustrates the need for coalition partners to integrate into the networks of their lead partners. The complexity of air to ground coordination was beyond the internal capability of British units to control such assets: 'It was widely acknowledged that had UK land forces received air support in greater quantities during TELIC than they did, they would have lacked the capability to control it without the assistance provided by the USMC Air Support Elements.'[72]

The joint issues of fire control, coordination of air and ground targeting, and the movement of ground forces gains additional administrative, technological, and political complexity in the 'combined' environment of a coalition battlespace. Intra-coalition fratricide has come to be one of the most politically sensitive areas governing the relations between partner nations, especially given that no coalition nation can completely provide their own close air support over their own troops, and most cannot provide any. As such, coalition partners rely heavily on American close air support. Administratively, this means coalition partners must be well practised in American fire control procedures to both request air support/indirect fire and protect their ground forces from fratricide. More specifically, in order to meet the demands of the American system of planning, coalition partners must be able to interoperate with American technical systems controlling the assignment, direction, and command of such assets. Even so, coalition forces dependent on American support may lack all that they require for the simple reason that American forces generally have priority in the assignment of scarce resources. Ground forces incapable of meeting these administrative and technological demands simply cannot operate safely on the modern digital battlefield. Again, British units were able to take advantage of privileged professional and political relationships despite their relative lack of technological sophistication; however, other countries cannot count on similar levels of support.[73]

Fratricide: limited interest and technological requirements in coalitions

The technical/administrative issues discussed above point to the one fracture line that underlies all coalition operations, especially those in dangerous ground situations. As military capability of America's partners, including that of its largest ones like the UK, shrinks, the political interest of states committing to military ventures shrinks correspondingly: with less ability to influence the strategic outcome, states invest less political interest in particular conflicts. As such, the relative political importance of casualties, especially those caused by fratricide, increases within a coalition. While America may be willing to (just) tolerate casualty figures in terms of hundreds of dead and wounded soldiers per month, similar figures, even reduced in terms of the scale of the military commitment, would prove politically ruinous for most coalition partners' governments. The fear of casualties places significant operational restrictions on what coalition partners will permit their ground forces to do once deployed.

Fratricide is in many ways a corollary to the issue of collateral damage. Just as the increasing precision of standoff weaponry has further spurred demands for the conduct of scrupulously clean targeting, the need to reduce the risk associated with fratricide within coalitions will predicate a similar demand for precision. The fallout from such a development, of course, will be increasing reliance on American systems given the relative lack of capability most coalition forces will bring in this area, especially in terms of air to ground resources. In other words, technology becomes increasingly important to reduce the level of political risk within operations assumed by military commanders and the planners under their control.[74]

Reducing the risk of fratricide is not completely reliant on technological solutions: pre-deployment training in 'tactics, techniques, and procedures' (TTP) can alleviate the risk to a considerable degree.[75] However, as TTP become increasingly reliant on digital systems, there is a real possibility of a fundamental disjuncture opening up between those using new digital systems and those relying, wholly or partially, on older analogue systems.

> Success in combat requires commanders to visualise the battlefield and to see the enemy, the terrain, and themselves in time and space. That, in a nutshell, is the point of all the rhetoric of information dominance. However, this has been difficult for higher echelons like a corps because of the time required to receive and process operations reports (information about what friendly forces are doing) and intelligence reports (information about what the enemy is doing).[76]

As discussed, new technology has aided considerably with this traditional military problem. Even as it has, long-range precision weaponry has increased the complexity of this visualisation process by widening the area of responsibility assumed by military commanders. This process has quite naturally been accompanied by a concomitant increase in terms of the information requirements, not only to target but also to manage the effects of such weaponry. Only systems capable of providing such information can be permitted to operate in the coalition battlespace given the risk of fratricide and collateral damage.[77] This in turn, has driven growing reliance on American systems.

In many regards, the problems outlined above are very traditional coalition and alliance issues. Geographical separation between forces of differing nationalities has long been used as a solution to interoperability mismatches and as a technique for reducing the possibility of fratricide. In Iraqi Freedom, the introduction of technology like FBCB2 seemed to indicate that such problems might be alleviated for the first time by giving all military commanders equal ability to visualise the battlespace and thus more efficiently coordinate their activities in time and space. However, as even the OFT-sponsored case study examining UK/US operations in OIF concluded, there was 'no evidence' of this as 'US and UK units appeared to operate as separate Division/Brigades at the operational and tactical levels'.[78]

Technology that was meant to aid in reducing and eliminating the friction of the battlefield, in coalition environments, may result in increasing it significantly. In a Rand Arroyo Center study of the impact of US Army transformation plans on coalition compatibility, a whole series of problems were projected to result from technology mismatches between coalition partners, including compromised American combat and combat service capabilities, increased fratricide, lowered operational tempo, compromised force protection, decreased logistics inefficiencies, and reduced overall situational awareness stemming from an inability to share operational data or to exploit fully available resources.[79] Each of these problems lessened the attractiveness of using coalition partners in operations given the problems that accompany their presence on the battlefield. Technological incompatibilities can be solved through 'fixes', generally established from sharing technology of the development of long-term policies that 'bridge' the gaps between two nations' technical systems with non-technical means. More typically, however, work-arounds are developed between partners. These are solutions that seek to reduce the impact of incompatibilities rather than eliminating them entirely.[80] Finally, the use of liaison teams, inserted as far down organisationally into a partner's land force as resources permit, can also lessen the impact of force incompatibility.[81] The key to any of these solutions working, however, is sufficient time to train with the technology, doctrine, and any associated work-around procedures. In operations demanding rapid response, the lack of time to prepare properly for the mission may place extraordinary pressures to rely on those partners that already have extensive experience operating with the US or to forego coalition participation altogether.[82]

In 2006, Kenneth Kreig, Undersecretary of Defense for Acquisition, Technology, and Logistics speaking at a defence industry conference noted that 'Our futures depend on international cooperation', without which, referring to America's partners, he noted 'I do not think we will be able to keep pace or benefit from each others advances'. Ominously, he concluded that there was 'no magic interoperability pill', however.[83] The absence of an 'interoperability black box'[84] points to the complex interrelationship of technology, doctrine, and even culture in the prosecution of military missions. Such interrelationships make the problem of interoperability resistant to simple technological solutions. Layered on top of these already pressing problems is the issue of national security which adds the additional complexity of restrictions in sharing information. Last of all, is the growing cost of military technology which both slows the progress of its development and implementation. All of these issues challenge the ability of nations to keep pace with the United States.

While the disparity between the US and the rest of the world is partly a product of its own design in order to reinforce its unipolar military status, it in turn creates growing problems for its own operations. Burden sharing during the Cold War was a hypothetical issue that considered the implications of an East–West war fought in Central Europe. The War on Terror is an ongoing operational drain on the forces of all who are participating in it, none the more so than the armed forces of the US which are experiencing a significant drain on their capacity to regenerate and continue the current fight as well as prepare for other global contingencies. Coalition partners are critical in this venture. However, the same resource challenges that beset American operations affect its partners even more. None are in a political or economic position to increase their effort, and the expense of operations make investment in developing future capital projects to catch up to the US even more difficult. The political fall out from either fratricide or collateral damage drive greater and greater reliance on US technology and military forces.

One observer noted:

> Units without digital battle command capability are non-networked forces in a networked world. Attempting to operate without the situational awareness the network provides makes units 'invisible' to others in the network; they can quickly become a danger to themselves and others – a liability rather than a strength.[85]

As ground forces expand the size of the physical area in which they operate, the area outside of their technological influence will shrink. This has largely already happened in the air, where American forces dominate the entire battlespace and coalition units must integrate as best they can within the technological and administrative parameters governing the use of air power. On the land, this point may also be approaching, where the geographical footprint of American land power expands to where there is no more available space for forces incapable of integrating into the American operating system. At that point, geographical separation as a tool of managing interoperability friction will no longer be available – in order to participate, total integration will be required.

Conclusion

The nature of warfare is changing in ways that are difficult to anticipate completely. On the one hand, the complexity of high-tech warfare as conducted by the United States is increasing at a rate the rest of the world cannot match. The best that other states can hope for is the pursuit of advances in niche areas of specialisation. At the same time, we are seeing the growing effect of insurgents and terrorists around the world. This is a natural response to the overwhelming superiority that economically advanced states' militaries, especially America's, enjoy in conventional warfare.

Despite this uncertainty, the modernisation plans of the US armed forces are well in place, and are determining how other developed nations approach the practice of war in a manner unprecedented in history. NCW, originally developed to take operational advantage of the concatenation of military sensors developed through the long Cold War, is now influencing Western military thinking. While setbacks in Iraq and Afghanistan may have tempered some of the initial enthusiasm for the strategic impact networks were supposed to have, IT continues to play significant roles in shaping tactical and operational engagements in both conflicts. If smaller powers wish to have any part in the military operations influencing the current strategic environment, they must seek greater interoperability with America's new military operating system. If they do not do so, they risk irrelevancy.

But the obstacles to full engagement in US-led NCW are not only technological. There are major political impediments as well. There is a fundamental disjuncture between the military and the political environment. As Western militaries pursue their visions of digital operations, the political realm remains imbued with an inherently subjective and nuanced nature that fails to translate into a digitised environment. This is most evident in coalition operations. At the heart of every alliance and – especially – coalition there is a tension between political strategy and military effectiveness. This tension is resolved only through compromises arrived at by hard negotiation. The digital logic of machines cannot recognise such human arrangements.

NCW offers clear advantages to militaries. The speed, precision, and reach of networked militaries make them difficult to counter on the battlefield. These advantages come with a clear price, however. The information that makes these forces so deadly must be carefully protected from damage and disclosure. The contemporary 'coalition of the willing' environment means that today's partner will look on from the sidelines tomorrow, and could even oppose a future coalition. In this regard, information release policy is not only an immediate operational concern, but also an issue of longer-term strategic importance. Digital protocols guarding information security can never be as flexible and malleable as human agreements. With an eye on the long term rather than what is immediately expedient, these protocols will always be more robust than what the situation might require.

Thus, there is a triangular relationship between NCW, information release policy, and coalition strategy. NCW aims for perfectly efficient military operations that alleviate the problems of operational choice in a confusing setting; the price is an environment of trust that permits free creative activity. Coalition strategies seek to increase political legitimacy or military resources; the price is political compromise between the differing plans of coalition partners. Finally, information security ultimately seeks to guard national security; the price is tight control.

As the case of naval operations in the Persian Gulf demonstrates, networks can be used successfully within coalition environments. Networking technologies were crucial to Australian-led operations in the northern Persian Gulf. Canadian-led operations in the Gulf of Oman used networking technologies to maintain a fragile coalition in a mission that was an important component of the War on Terror. The freedom permitted these two coalition partners, however, was decisively dependent on the US Navy's trust in the Australian and Canadian navies. Such cooperation was generated by a particular relationship between trust, security, and compromise, specific to a particular time, place, and group of participants. These factors will not necessarily be present in other circumstances.

In fact, the case of NORAD shows the limits that even high levels of professional trust between military services has on enhancing cooperation between close political partners. NORAD arguably practised a form of network centric warfare in its concern over the air defence of North America. Canadian and American operational plans to counter any Soviet attack were contingent on information sharing from sensors based across the North American continent. This close cooperative arrangement was viewed with alarm on both sides of the border, however; Canadian politicians and American military officers were equally concerned for the sovereign control of bi-national operations. Paradoxically, as the integration of each nation's economic and civil infrastructure proceeds unabated, the level of military integration levelled off early in the Cold War and has been progressively ratcheted back since. Arguably, today there is less integration within NORAD itself than in the 1950s when the regional and sector control boundaries spanned the Canada–US border. Further, with the post-9/11 security challenge becoming increasingly complex, the level of integration between the operational plans of each nation has also become more restricted. The digitisation of data has complicated this process, given the difficulties of sharing domestic security and law enforcement data across borders. While the professional trust shared between the US and Canadian air forces remains high, this has not been sufficient to permit greater levels of information sharing in other areas of the Canada–US security relationship.

On land we see, perhaps, the highest levels of complexity in the efforts to network military forces, as well as share information. The nature of the land environment is far different from that experienced by air and naval forces. The technical challenges of networking the exponentially larger organisations that armies represent has not been solved as yet. How far down the network reaches and how much information needs to be pushed or at least made available for consumption by soldiers are serious questions that remain to be solved.

The dilemmas presented by land operations go much deeper than technical and engineering problems, however. The land is infused with political significance in ways that are subtly different than at sea or in the air. We live on the land and thus we invest political, economic, and perhaps most importantly, cultural significance into the land in ways that are different at sea or in the air. These factors merge with the physical geography of an area to enhance the battlefield's complexity. The use of terrain by insurgent and terrorist movements, as well as their ability to disappear into civil society illustrates this issue in particular. Perspective and judgement are central to resolving these issues, however, it is these subjective assessments that make them irreducibly human activities, as opposed to automated digital processes. Furthermore, judgement is fundamentally impacted by political interest within coalition settings.

Networks and IT have obviously enhanced the ability of armies to move and fight, as evidence from Iraq clearly shows. The impact of technology like FBCB2 to enhance the sharing of battlefield data for operational effect cannot be denied. Again, however, the same technology that enhances an army's operations could be used offensively against it. Sharing FBCB2 data with unreliable forces poses extreme operational risks, as it effectively reveals not only the location of individual units, but also their status, orders, and intent. Such data will likely remain highly restricted amongst coalition partners, possibly shared only amongst the closest of collaborators.

Finally, technologies like FCS permit highly dispersed and mobile operations of individually weak units, given their interdependent nature and ability to draw on resources not fully under their own control. The impact on the battlefield is the progressive growth of the span of control each unit exercises. These features all will make it increasingly difficult for armies to cooperate on the digital battlefield. The time-worn techniques of geographical separation, traditionally used by coalitions and alliances to avoid the kinds of interoperability problems that lead to fratricide, will become more and more difficult to implement as smaller and smaller units control larger and larger areas. Further, few nations will be able to invest to the same level of effort as the Americans, ensuring that its partners will always be racing to keep pace with them, usually unsuccessfully.

The closing days of the Bush Administration are marked by serious efforts to repair the damage caused by the burst of exuberance and self-confidence in American military power to resolve the problems of international security. Iran and North Korea have been handled far more sensitively by the United States as compared with Iraq, and American diplomacy has been careful to pay greater attention to the choices and plans sought by its strategic partners in these areas. Furthermore, it is unlikely that the next administration will pursue American interests as aggressively. Militarily, however, we can expect little change in how the United States will conduct its operations. While Iraq may have tempered some of the hopes regarding 'dominant battlespace knowledge', there is a fundamental trust that IT provides greater efficiency to the conduct of military operations. The integration of technologies like UAVs continues apace, and while under constant threat, the Future Combat System seems set to define the future of the American army if only because the large manpower-intensive armies of the Cold War are unaffordable, even by the Americans.

In the 1950s, struggling to deal with the impact of an equally disruptive technology, nuclear weapons, the US Army developed its doctrine of the Pentomic division. In many ways, the Pentomic division, with its small, highly mobile units, each reliant on non-organic supporting services, resembled the proposed Future Combat Systems. Moreover, there are some broad similarities here between how both information and nuclear weapons would affect the battlefield. Each required that military services fundamentally rethink how to conduct operations without history serving as a guide for best practice.

The Pentomic division was ultimately abandoned as it became apparent that nuclear weapons had relatively little utility on the battlefield, and none at all in counter-insurgencies. While possessed of tremendous political utility in maintaining strategic stability between the great powers, and in manipulating crisis behaviour, nuclear weapons were essentially unusable militarily. Information is clearly different in this regard. As each case has highlighted, information *is* of tremendous use to military commanders whether that is in maintaining situational awareness, directing time sensitive targeting, or coordinating the movement of dispersed movement of tactical units over enormous distances. All this suggests that IT will not fade into the military background in the manner that tactical nuclear weapons did in the late 1960s and 1970s. However, as should be apparent, it is not clear that the doctrines that have evolved to accommodate the capabilities that IT introduces to the battlefield are the correct ones. In this regard, the requisite historical analogy may be more like the conundrums posed by the introduction of armour and mobility at the close of the First World War rather than nuclear weapons. There too, military planners faced considerable uncertainty and a wide variety of options in how to implement new technology. As the opening campaigns of the Second World War showed, some nations did a better job than others in this task.

Militaries are probably facing a similar task at the moment in accommodating the impact of IT in their doctrines. On the surface, the US clearly has been able to use IT effectively in battle to tremendous advantage. While the dash up the Euphrates remains controversial in terms of the number of forces assigned, and may have ultimately contributed to the subsequent slide into chaos, the nature of the advance would not have been possible in 1991. Furthermore, the United States has been able to introduce a variety of technologies that have been able to keep rough pace with the slew of innovative applications appearing on the Internet. It has done this at great cost, however. Curiously, much of the innovation that has appeared on the Internet has cost relatively little. Web 2.0 applications have been typically developed by individuals without enormous research and development budgets. Whether the United States can keep pace with the Internet hothouse remains to be seen. Clearly, however, America's military partners will be tremendously stressed by these developments, to say nothing about the information sharing difficulties that have been central to this analysis.

To return to the nuclear weapons analogy, the role of information on the battlefield remains equally problematic for the future. The opportunities and hazards that globalisation will generate in the coming decades will ultimately require action at the state level, and most likely, will require some form of military response. The ability for militaries to organise combined action will be an important factor in the success of these ventures. The impact of digital technology may play a pernicious role in restricting the ability of militaries to cooperate. On the surface, networks are seemingly ideal technology in this period of coalitions of the willing. Their ability to add and subtract nodes without affecting their overall structure is tailored precisely to the nature of come as you are coalition organisation. However, the lingering impact of national security means that information on military networks will flow in a significantly different manner than on social networks.

It may be that we are at a fundamental intersection between civilisational ages, between the centuries old industrialism that gave rise to nation states and their associated armies, and Castell's informationalism. How this divide will be bridged remains to be seen. While the social structures of the previous age are all under constant challenge from the forces of informationalism, the social networks of the Web will not solve the security challenges that will emerge from the opportunities and hazards of globalisation, these will all require the resources, organisation, and political legitimacy that only states provide. In the same way, we see how networked insurgents present real tactical conundrums to conventional forces, enhancing their power. Still, the strategic record of guerrilla movements is not one of dramatic success. Networks have clearly enhanced their power, giving them greater global reach and persistence, however, it is noteworthy that no military has opted to adopt their organisational structure. And yet, if informationalism is a real social process underway, as opposed to simply an abstract construction imposed by social scientists, then there will be obvious consequences for how militaries organise themselves and conduct their operations. In this, like the Pentomic division, network centric warfare may simply represent a first step in coming to grips with these challenges.

There will be immediate consequences, in the meantime. The foremost of these is the ever present search for interoperability amongst friendly military forces. The nature of NCW as a military operating system is driving this effort. However, we should not expect that these efforts will ultimately amount to much more than the sharing of unclassified data. The exception to this rule will be for extremely close partners. In this, the ABCA nations may have no other equivalent. The other likely consequence, thus, is the growing probability of unilateralism in the execution of military operations. Australian defence analyst Alan Ryan has noted that 'operational success in the twenty-first century operations will be the product of orchestrating the combat multiplier effects inherent in multinational forces. To achieve this effect is undoubtedly the acme of skill.'[1] These are words of hope, but they essentially reflect the experience of trusted partners close to the United States, like Australia, Canada, and the UK. For more distant security partners, the operational demands of information security will threaten military interoperability, and thus strategic cooperation. In twenty-first century operations, the US armed forces may increasingly 'go it alone'.

Notes

Introduction

1 Thomas Valovic, *Digital Mythologies: The Hidden Complexities of the Internet*, Piscataway, NJ: Rutgers University Press, 2000, p. 15.
2 Howard Rheingold, *Smart Mobs: The Next Social Revolution*, New York: Basic Books, 2002, p. 48.
3 Don Tapscott and Anthony D. Williams, *Wikinomics: How Mass Collaboration Changes Everything*, New York: Penguin Group, 2006, p. 1.
4 Frances Cairncross, *The Death of Distance*, Boston: Harvard Business School Press, 1997.
5 Thomas Friedman, *The World is Flat*, New York: Farrar, Straus and Giroux, 2005.
6 Vincent Moscoe, *The Digital Sublime: Myth, Power, and Cyberspace*, Cambridge MA: MIT Press, 2004.
7 Manuel Castells, *The Rise of the Network Society, Second Ed.*, Malden MA: Blackwell, 2000, pp. 508–509.
8 Christopher Coker's work does begin to address some of the social implications that new forms of technology are raising for military forces. See Christopher Coker, *Waging Wars Without Warriors: The Changing Culture of Military Conflict*, London: IISS, 2002 and *The Future of War: The Re-enchantment of War in the Twenty-first Century*, London: Blackwell Manifestos, 2004.
9 One example of a systems level approach is the body of research examining the use of networked entities to confront hierarchically organised foes. This includes John Robb's *Brave New War: The Next Stage of Terrorism and the End of Globalization*, Hoboken, NJ: John Wiley and Sons, 2007, as well as a burgeoning literature on 'swarms' and 'fourth generation warfare'.
10 David Schmidtchen, *The Rise of the Strategic Private: Technology, Control and Change in a Network Enabled Military*, Duntroon ACT: Land Warfare Studies Centre, 2006, pp. 27, 43.
11 Manuel Castells notes:

> Technology per se does not determine historical evolution and social change, technology (or the lack of it) embodies the capacity of societies to transform themselves as well as the uses to which societies, always in a conflictive process, decide to put their technological potential.
>
> (Castells, *The Rise of the Network Society*, pp. 7–13)

12 This argument is developed most fully over Castells' three volume masterpiece *The Information Age: Economy, Society and Culture*, Malden MA: Blackwell, 1996.
13 Manuel Castells, 'The Network Society', in Pekka Himanen (ed.), *The Hacker Ethic*, New York: Random House, 2001, pp. 156–158.

14 Nico Stehr, 'A World Made of Knowledge', available at www.inco.hu/inco0401/global/cikk1h.htm.
15 Valovic, *Digital Mythologies*, p. 25.
16 Stehr, 'A World Made of Knowledge'; Castells, *The Rise of the Network Society*, p. 158.
17 Valovic, *Digital Mythologies*, p. 23.
18 Scott Lash, *Critique of Information*, London: Sage Publications, 2002, pp. 140–145.
19 Nicholas Negroponte, *Being Digital*, New York: Vintage Books, 1995, p. 8.
20 Castells, *The Rise of the Network Society*, p. 31.
21 David Weinberger, *Loosely Joined Pieces: A Unified Theory of the Web*, Cambridge MA: Perseus Publications, 2002, pp. ix–xi.
22 Castells, *The Rise of the Network Society*, p. 23.
23 Jerry Everard, *Virtual States: The Internet and the Boundaries of the Nation-State*, London: Routledge, 2000, p. 46.
24 Don Tapscott and Anthony D. Williams, *Wikinomics*, p. 37. A good overview of this fast evolving space is found in 'The Future of Web 2.0', *Technology Review*, July/August 2008, pp. 34–69.
25 Howard Rheingold, *Smart Mobs: The Next Social Revolution*, pp. 60–61.
26 Tapscott and Williams, *Wikinomics*, p. 271.
27 Castells, *The Rise of the Network Society*, pp. 501–502.
28 Castells, 'The Network Society', pp. 166–167.
29 Tapscott, Williams, *Wikinomics*, p. 11.
30 Howard Rheingold, *Smart Mobs: The Next Social Revolution*, p. 39; Eric S. Raymond, 'The Cathedral and the Bazaar', www.catb.org/~esr/writings/cathedral-bazaar/ cathedral-bazaar/, p. 22.
31 Tapscott and Williams, *Wikinomics*, pp. 20–30.
32 Raymond, 'The Cathedral and the Bazaar', p. 23.
33 Richard Stallman, 'Free Software: Freedom and Cooperation', 29 May 2001, www.gnu.org/events/rms-nyu-2001-transcript.html.
34 Chris Anderson, 'Free! Why $0.00 Is the Future of Business', *Wired*, 25 February 2008, www.wired.com/techbiz/it/magazine/16–03/ff_free.
35 John Perry Barlow, 'The Economy of Ideas', *Wired*, 2 March 1994, www.wired.com/wired/archive/2.03/economy.ideas.html.
36 Lash, *Critique of Information*, p. 159.
37 Raymond, 'The Cathedral and the Bazaar', pp. 8–9.
38 James Kinniurgh and Dorothy Denning, 'Blogs and Military Information Strategy', *Joint Special Operations University Report 06–5*, June 2006, pp. 1–2.
39 Barlow, 'The Economy of Ideas'.
40 Lash, *Critique of Information*, p. 149.
41 Barlow, 'The Economy of Ideas'.
42 Stallman, 'Free Software: Freedom and Cooperation'.
43 Raymond, 'The Cathedral and the Bazaar', p. 23.
44 Derek S. Reveron, 'Old Allies, New Friends: Intelligence Sharing in the War on Terror', *Orbis*, Summer 2006, p. 453.
45 Clive Thompson, 'Open Source Spying', *New York Times Magazine*, 3 December 2006, www.nytimes.com/2006/12/03/magazine/03intelligence.html; Calvin Andrus, 'The Wiki and the Blog: Toward an Adaptive Intelligence Community', *Studies in Intelligence*, vol. 49, no. 3, September 2005, available at http://ssrn.com/abstract=755904.
46 Scott Shane, 'Logged in and Sharing Gossip, er, Intelligence', *New York Times*, 2 September 2007, www.nytimes.com/2007/09/02/weekinreview/02shane.html.
47 Barlow, 'The Economy of Ideas', *Wired*.
48 Anderson, 'Free! Why $0.00 Is the Future of Business'.
49 Richard Best, 'Sharing Law Enforcement and Intelligence Information: The Congressional Role', *Congressional Research Service Paper RL33873*, 13 February 2007.
50 Cited in Tapscott and Williams, *Wikinomics*, p. 17.

51 Jaron Lanier, 'Digital Maoism: The Hazards of New Online Collectivism', *Edge: The Third Culture*, 30 May 2006, available at www.edge.org/3rd_culture/lanier06/lanier06_index.html.
52 Nikolai Bezroukov, 'A Second Look at the Cathedral and the Bazaar', *First Monday*, vol. 4, no. 12, December 1999, www.uic.edu/htbin/cgiwrap/bin/ojs/index.php/fm/article/view/708/618.
53 Tapscott and Williams, *Wikinomics*, p. 25.
54 Ibid., p. 25.
55 *Feasibility Report: Report for the Congress of the United States, March 2008*, prepared by the Program Manager, Information Sharing Environment, p. 14, available at www.fas.org/irp/agency/ise/feasibility.pdf.
56 Ibid., p. 17.
57 Bezroukov, 'A Second Look at the Cathedral and the Bazaar'.
58 Nicky Hager, *Secret Power: New Zealand's Role in the International Spy Network*, Nelson, NZ: Craig Potton Publishing, 1996, pp. 23–24.
59 Steven Cambone, 'Memorandum: Security Classification Marking Instructions', 27 September 2004.
60 James Kinniurgh and Dorothy Denning, 'Blogs and Military Information Strategy', *Joint Special Operations University Report 06–5*, June 2006, p. 3.
61 See, for example, SECNAV INSTRUCTION 5720.47B 'Department of the Navy Policy for Content on Publically Accessible World Wide Web Sites', 28 December 2005; Army Regulation 530–1 'Operations and Signal Security', 19 April 2007, paragraph 2–1 g; available at http://blog.wired.com/defense/files/army_reg_530_1_updated.pdf.
62 See, for example, http://destroyermen.blogspot.com/2008/04/please-stand-by.html and the subsequent post http://destroyermen.blogspot.com/2008/04/top-cover.html.
63 Emphasis added. Army Regulation 530–1 'Operations and Signal Security', 19 April 2007, p. 1.
64 See Alice R. Buchalter, John Gibbs and Marieke Lewis, *Laws and Regulations Governing the Protection of Sensitive but Unclassified Information*, Washington DC: Library of Congress, September 2004. Available at www.loc.gov/rr/frd/pdf-files/sbu.pdf.
65 Stehr, 'A World Made of Knowledge'.
66 Executive Order 12958 governs the modes of classifying national security information: www.fas.org/irp/offdocs/eo12958.htm. Davi M. D'Agostino, *Managing Sensitive Information: DoD Can More Effectively Reduce the Risk of Classification Errors*, Washington DC: Government Accounting Office, June 2006, p. 2.
67 Davi M. D'Agostino, *Managing Sensitive Information*, pp. 4–5.
68 Lash, *Critique of Information*, p. 146.
69 Derek S. Reveron, 'Old Allies, New Friends: Intelligence Sharing in the War on Terror', *Orbis*, Summer 2006, p. 459.
70 Gene I. Rochlin, *Trapped in the Net: the Unanticipated Consequences of Computerization*, Princeton: Princeton University Press, 1997, p. 9.
71 For example, Jonathon Zittrain, *The Future of the Internet – And How to Stop It*, New Haven, CT: Yale University Press, 2008; Ron Deibert, John Palfrey, Rafal Rohinsky, and Jonathon Zittrain, *Access Denied: The Practice and Policy of Global Internet Filtering*, Boston: Harvard University Press, 2008.
72 Manuel Castells, *End of the Millenium*, Malden MA: Blackwell, 1998, p. 371.
73 Valovic, *Digital Mythologies*, p. 21. The lack of civility in digital communication has been remarked on by many: Alan Jacobs, 'Goodbye, Blog', www.christianitytoday.com/global/printer.html?/bc/2006/003/17.36.html.
74 Jerry Everard, *Virtual States: The Internet and the Boundaries of the Nation*-State, London: Routledge, 2000, p. 44.
75 'The proposed network is a *universal high secrecy system* made up of a hierarchy of *less secure subsystems*. It is proposed that the network will treat all inputs as if they are classified in order to increase the intercept price to the enemy to a value so high

that interception would not be worth his efforts.' Emphasis in the original. Paul Baran, *On Distributed Communications: IX. Security, Secrecy, and Tamper Free Considerations*, Santa Monica: Rand Corporation, 1964, p. 7.

76 Castells, *End of the Millenium*, pp. 376–377.

1 US military primacy and the new operating system

1 See www.globalsecurity.org/military/world/spending.htm.

2 Ann Scott Tyson, 'Military Goals Claim Priority over Diplomacy', *Christian Science Monitor*, vol. 93, no. 231, 24 October 2001, p. 3; 'In Rumsfeld's Words: Guidelines for Committing Forces', *New York Times*, 14 October 2002, p. A9.

3 As Victor David Hanson and David B. Ralston have both argued, the creeping standardisation of military practice is nothing new. However, in the past other powers have had several models from which to choose. Japan, for example, modelled its navy on the British example and its army on the German. At present, all look to the American military for guidance on doctrinal policy and capital investment. Victor David Hanson, *Carnage and Culture: Landmark Battles in the Rise of Western Power*, New York: Doubleday, 2001; David B. Ralston, *Importing the European Army*, Chicago, IL: University of Chicago Press, 1990.

4 Of course, this does not imply that the United States is militarily omnipotent, as the ongoing insurgencies around the world amply demonstrate. But as Hanson points out, where smaller powers challenge the US, they do so in their own lands, and often use technology developed by the United States. These powers have not been able to develop indigenous technology capable of defeating the US, nor are they free to operate in the heartland of North America. Hanson, *Carnage and Culture*, pp. 443, 453.

5 Barry Posen, 'Command of the Commons', *International Security*, vol. 28, no. 1, 2003, pp. 8–9.

6 Ibid., pp. 17–18.

7 The first *National Security Strategy*, in 1991, lists a range of 'Interests and Objectives' that the US would pursue 'in concert with its allies'. In the 2002 version, it is noted that the US will 'Strengthen alliances to defeat Global Terrorism and work to prevent attacks against us and our friends'. See www.fas.org/man/docs/91805-nss.htm; www.whitehouse.gov/nsc/nss3.html.

8 Kenneth Waltz, 'Globalization and American Power', *The National Interest*, Spring 2000, p. 54.

9 Robert Jervis, *American Foreign Policy in a New Era*, New York: Routledge, 2005, p. 12.

10 Ibid., p. 31.

11 Christopher Layne, 'America as European Hegemon', *The National Interest*, Summer 2003, p. 28.

12 Raymond Aron, *Peace and War: A Theory of International Relations*, New York: Doubleday, 1966, p. 99.

13 Joseph Nye, 'Military De-Globalization', *Foreign Policy*, January–February 2001, pp. 82–83.

14 David Calleo, 'Power, Wealth, and Wisdom: The United States and Europe after Iraq', *The National Interest*, Summer 2003, p. 12.

15 Layne, 'America as European Hegemon'.

16 Barton Gellman, 'Pentagon Would Preclude a Rival Superpower', *Washington Post*, 11 March 1992, p. A1.

17 *Interview: Dennis Ross*, www.pbs.org/wgbh/pages/frontline/shows/iraq/interviews/ross.html.

18 *Interview: Barton Gellman*, www.pbs.org/wgbh/pages/frontline/shows/iraq/interviews/gellman.html.

19 *Excerpts From 1992 Draft 'Defense Planning Guidance'*, http://www/pbs.org/wgbh/pages/frontline/shows/iraq/etc/wolf.html.
20 Calleo, 'Power, Wealth, and Wisdom', p. 7.
21 *Excerpts From 1992 Draft 'Defense Planning Guidance'*.
22 For example:

> It has taken almost a decade for us to comprehend the true nature of this new threat. Given the goals of rogue states and terrorists, the United States can no longer solely rely on a reactive posture as we have in the past. The inability to deter a potential attacker, the immediacy of today's threats, and the magnitude of potential harm that could be caused by our adversaries' choice of weapons, do not permit that option. We cannot let our enemies strike first. ... The United States has long maintained the option of preemptive actions to counter a sufficient threat to our national security. The greater the threat, the greater is the risk of inaction – and the more compelling the case for taking anticipatory action to defend ourselves, even if uncertainty remains as to the time and place of the enemy's attack. To forestall or prevent such hostile acts by our adversaries, the United States will, if necessary, act preemptively.
>
> (*The National Security Strategy of the United States*, 2002, www.whitehouse.gov/nsc/nss5.html)

23 Quoted in Jervis, *American Foreign Policy*, p. 90.
24 Hubert Védrine famously described the United States as a 'hyper-power' (*hyper puissance*) during the Clinton administration. Calleo, 'Power, Wealth, and Wisdom', p. 8.
25 John Shalikashvili, *Joint Vision 2010*, Washington DC: Joint Chiefs of Staff, 1997, p. 25.
26 Department of Defense, *Transformation Planning Guidance*, April 2003, p. 3.
27 Paul Wolfowitz, 'Remembering the Future', *The National Interest*, Spring 2000, p. 41.
28 Robert Kagan and William Kristol, 'The Present Danger', *The National Interest*, Spring 2000, p. 63.
29 Robert B. Strassler, *The Landmark Thucydides*, New York: The Free Press, 1996, p. 114.
30 Kishore Mahbubani, 'The Impending Demise of the Postwar System', *Survival*, vol. 47, no. 4, Winter 2005–2006, p. 17.
31 Walter Russell Mead, *Power Terror and War: American Grand Strategy in a World at Risk*, New York: Alfred A. Knopf, 2004, p. 120. From a different political perspective, Christopher Layne agrees with this conclusion. 'The damage inflicted on Washington's ties to Europe by the Bush Administration is likely to prove real, lasting, and, at the end of the day, irreparable.' Layne, 'America as European Hegemon', p. 17.
32 Robert E. Osgood, *The Entangling Alliance*, Chicago, IL: University of Chicago Press, 1962, p. vii; Henry Kissinger, *The Troubled Partnership*, New York: McGraw Hill, 1965, p. 5.
33 Ian Clark, *Globalization and Fragmentation: International Relations in the Twentieth Century*, Oxford: Oxford University Press, 1997, p. 1; R. J. Barry Jones, *Globalisation and Interdependence in the International Political Economy: Rhetoric and Reality*, London: Pinter Publishers, 1995, p. 13; Andrew Hurrell, 'Explaining the Resurgence of Regionalism in World Politics', *Review of International Studies*, vol. 21, no. 4, 1995, p. 345.
34 Christopher Coker, *Globalisation and Insecurity in the Twenty-first Century: NATO and the Management of Risk, Adelphi Paper #345*, London: IISS, 2002, p. 21; Clark, *Globalization and Fragmentation*, p. 18.
35 Coker, *Globalisation and Insecurity*, p. 25; Mary Kaldor, *New and Old Wars: Organised Violence in the Global Era*, Cambridge: Polity Press, 1999, p. 70; M. Singer and A. Wildavsky, *The Real World Order: Zones of Peace/Zones of Turmoil*, Chatham, NJ: Chatham House, 1993, pp. 4, 6; Thomas P. Barnett, *The Pentagon's New Map: War and Peace in the Twenty-First Century*, New York: Berkley Books, 2004.

36 Frank G. Hoffman, 'The New Normalcy', *E-Notes*, www.fpri.org/enotes/20060512.americawar.hoffman.newnormalcy.html; Charles C. Krulak, 'The Strategic Corporal: Leadership in the Three Block War', *Marines Magazine*, January 1999.
37 Coker, *Globalisation and Insecurity*, pp. 27–31.
38 Moisés Naim, 'The YouTube Effect', *Foreign Policy*, January/February 2007; Claude Salhani, 'Cell Phone Cams Exposing Torture', *SpaceWar Daily*, 16 January 2007. www.spacewar.com/reports/Cell_Phone_Cams_Exposing_Torture_999.html.
39 Naim, ibid., See also Greg Sandoval, 'Now Playing on the Net: War Propaganda', CNET News.com, 22 August 2006, http://news.cnet.com/Now-playing-on-the-Net-War-propaganda/2100–1038_3–6108004.html.
40 Maria Aspan, 'Ease of Alteration Creates Woes for Picture Editors', *New York Times*, 14 August 2006, www.nytimes.com/2006/08/14/technology/14photoshop.html; Michelle Malkin, 'The Photo Op Shop of Horrors', *Washington Times*, 19 August 2006, www.washingtontimes.com/news/2006/aug/18/20060818–091848–7126r/.
41 Lawrence Freedman, 'The Transatlantic Agenda: Vision and Counter-Vision', *Survival*, vol. 47, no. 4, Winter 2005–2006, p. 20.
42 Ulrich Beck, *Risk Society: Towards a New Modernity*, London: Sage, 1992; Ulrich Beck, *World Risk Society*, Cambridge: Polity Press, 1999; Anthony Giddens, *The Consequences of Modernity*, Cambridge: Polity, 1990; J. Franklin (ed.), *The Politics of Risk Society*, Cambridge: Polity, 1998; Barbara Adam, Ulrich Beck, and Joost van Loon, *Risk Society and Beyond: Critical Issues for Social Theory*, London: Sage, 2000.
43 Coker, *Globalisation and Insecurity*, p. 57; Ulrich Beck, *Risk Society: Towards a New Modernity*, p. 2.
44 Anthony Giddens, *Runaway World: How Globalisation is Reshaping our Lives*, London: Profile Books, 1999, p. 26.
45 Beck, *Risk Society*, p. 29.
46 Ibid., p. 13.
47 Scott Lash and Bryan Wynne, 'Forward', in Beck (ed.), *World Risk Society*, p. 4.
48 Beck, *Risk Society*, p. 27.
49 Coker, *Globalisation and Insecurity*, pp. 72–75.
50 Giddens, *Runaway World*, pp. 29–31.
51 Nico Stehr, 'A World Made of Knowledge', available at www.inco.hu/inco0401/global/cikk1h.htm.
52 Michael Ignatieff, *Virtual War: Kosovo and Beyond*, Toronto: Viking Press, 2000, p. 197.
53 Kathryn Cochrane, 'Kosovo Targeting – A Bureaucratic and Legal Nightmare', *Aerospace Centre Paper 3*, Canberra: Aerospace Development Centre, June 2001, p. 12.
54 Rosemary Foot, 'Introduction', in Rosemary Foot, John Lewis Gaddis, and Andrew Hurrell (eds), *Order and Justice in International Relations*, Oxford: Oxford University Press, 2003, p. 1.
55 Lawrence Freedman, 'Strategic Studies and the Problem of Power', in Lawrence Freedman, Paul Hayes and Robert O'Neill (eds), *War Strategy, and International Politics: Essays in Honour of Sir Michael Howard*, Oxford: Clarendon Press, 1992.
56 Andrew Hurrell, 'Order and Justice in International Relations: What Is at Stake?', in Foot, Gaddis, and Hurrell (eds), *Order and Justice in International Relations*, p. 27.
57 Ignatieff, *Virtual War*, p. 201.
58 In mid-2006, the US pledged $116 million at a Sudan donors' conference, the largest contribution of all the delegations present. 'United States Commits $116 Million at Sudan Donors Conference', State Department press release, 19 July 2006, www.state.gov/r/pa/prs/ps/2006/69224.htm.
59 Freedman, 'Strategic Studies and the Problem of Power', p. 290.
60 Ignatieff, *Virtual War*, p. 203.
61 Bruce R. Nardulli *et al.*, *Disjointed War: Military Operations in Kosovo, 1999*, Santa Monica, CA: RAND Arroyo Center, 2002, p. 2.

62 William Shawcross, *Allies: The US, Britain, Europe and the War in Iraq*, London: Atlantic Books, 2003, pp. 82–83.
63 Cochrane, 'Kosovo Targeting', p. 13.
64 Ibid., p. 11.
65 Ignatieff, *Virtual War*, pp. 198–200.
66 Freedman, 'Strategic Studies and the Problem of Power', pp. 291–293.
67 Freedman, 'The Transatlantic Agenda', p. 30.
68 The adoption of Israeli urban-warfare techniques is a good example of this tendency. Justin Huggler, 'Israelis Trained US Troops in Jenin-Style Urban Warfare', *The Independent*, 29 March 2003.

2 Freedom and control: networks in military environments

1 Department of Defense, *Transformation Planning Guidance*, April 2003, p. 1.
2 Colin S. Gray, *Strategy for Chaos*, London: Frank Cass, 2002, pp. 13–17.
3 Eliot Cohen, 'Change and Transformation in Military Affairs', *Journal of Strategic Studies*, vol. 27, no. 3, September 2004.
4 See Williamson Murray, 'May 1940: Contingency and Fragility of the German RMA', in MacGregor Knox and Williamson Murray (eds), *The Dynamics of Military Revolution, 1300–2050*, Cambridge: Cambridge University Press, 2001; Thomas G. Mahnken, 'Beyond Blitzkreig: Allied Responses to Combined-Arms Armoured Warfare during World War II', in Emily O. Goldman and Leslie C. Eliason (eds), *The Diffusion of Military Technology and Ideas*, Stanford, CA: Stanford University Press, 2003; Williamson Murray, 'Armored Warfare: The British, French, and German Experiences', in Williamson Murray and Allan R. Millet (eds), *Military Innovation in the Interwar Period*, Cambridge: Cambridge University Press, 1996; Barry R. Posen, 'The Battles of 1940', *The Sources of Military Doctrine*, Ithaca, NY: Cornell University Press, 1984.
5 Cebrowski and Gartska, 'Network Centric Warfare', pp. 28–35.
6 In this respect one need only think of the Battle of Midway. Three-dimensional warfare presents far more complex command and control issues than the traditional naval battleline. Karl Lautenschlager, 'Technology and the Evolution of Naval Warfare', *International Security*, vol. 8, no. 2, 1983.
7 In 1942, Admiral Ernest J. King asked Vannevar Bush of the Office of Scientific Research and Development to examine the possible development of a system of radar relays that would permit ships to share radar information, thus increasing commanders' awareness of the tactical situation. The project later switched to a system of air-based radars, which ultimately saw the development of the first airborne early-warning aircraft in the form of modified Grumman *Avengers* carrying APS-20 radars. Edwin Leigh Armistead, *AWACS and Hawkeyes*, St Paul, MN: MBI Publishing, 2002, pp. 3–7.
8 In 1957, after three years of deliberation, the CANUKUS Naval Data Transmission Working Group ratified the technical standard for data exchange. Originally named the Tactical International Data Exchange (TIDE, 'good for cleaning up messy tactical pictures'), it later became known as Link 2 (given as 'II' in Roman numerals) in the Royal Navy, which was already using data-sharing technology to distribute tactical information among its ships. As other NATO links became established, Link II became known as 'Link 11'. Norman Friedman, *World Naval Weapons Systems 1997–1998*, Annapolis, MD: Naval Institute Press, 1997, p. 28.
9 Robert Burnett and P. David Marshall, *Web Theory: An Introduction*, London: Routledge, 2003, p. 25; Norbert Weiner, *Cybernetics; Or, Control and Communication in the Animal and the Machine*, New York: Wiley, 1948.
10 Tacticians anticipated that Soviet bombers would mass their aircraft in 'regimental' attacks, launching waves of missiles at naval formations in the hope of overwhelming

their defences. In this type of tactical environment, it would no longer be possible to coordinate the defence of a task force through voice reporting, nor could the resources of any single ship defend against such an attack. This meant that the area that had to come under positive control by Western ships and aircraft expanded considerably. Norman Friedman, *The US Maritime Strategy*, London: Jane's Publishing, 1988, pp. 162–164, 174; Scott L. Nicholas, 'Anti-carrier Warfare', in Bruce W. Watson and Susan M. Watson (eds), *The Soviet Navy: Strengths and Liabilities*, Boulder, CO: Westview, 1986, p. 146; Norman Friedman, *US Destroyers Revised Edition*, Arlington, VA: Naval Institute Press, 2004, pp. 391–392.

11 Jacob Neufeld, George M. Watson Jr, and David Chenoweth (eds), *Technology and the Air Force: A Retrospective Assessment*, Washington DC: USAF, 1997.

12 See, for example, Colonel Thomas A. Cardwell (USAF), *Airland Combat: An Organization for Joint Warfare*, Maxwell, AL: Air University Press, 1992, pp. 75–80; the concept of 'Agility', defined as 'the ability of friendly forces to act faster than the enemy' is clearly derived from Col. John Boyd's OODA loop. Dept. of the Army, *US Army Field Manual 100–5 Blueprint for the AirLand Battle*, Washington DC: Brassey's (US) Inc., 1991, pp. 16–17.

13 Norman Friedman, *The Fifty Year War: Conflict and Strategy in the Cold War*, Annapolis, MD: United States Naval Institute Press, 2000, pp. 445–451.

14 Milan Vego, *Operational Warfare*, Newport, RI: Naval War College, 2000, pp. 1–2.

15 Timothy Travers, *The Killing Ground: The British Army, the Western Front, and the Emergence of Modern Warfare, 1900–1918*, London: Allen Unwin, 1987; Murray, 'Armored Warfare'; Jonathan B. A. Bailey, 'The First World War and the Birth of Modern Warfare', in Knox and Murray (ed.), *The Dynamics of Military Revolution*; Gray, *Strategy for Chaos*.

16 Bailey, 'The First World War and the Birth of Modern Warfare', p. 132. Emphasis added.

17 Defined by the Department of Defense as: 'The environment, factors, and conditions that must be understood to successfully apply combat power, protect the force, or complete the mission. This includes the air, land, sea, space, and the included enemy and friendly forces; facilities; weather; terrain; the electromagnetic spectrum; and the information environment within the operational areas and areas of interest'. Department of Defense, *Joint Publication 1–02, 'DOD Dictionary of Military and Associated Terms'*, www.dtic.mil/doctrine/jel/doddict/data/b/00700.html, as amended through 31 August 2005.

18 Admiral William A. Owens, 'The Emerging System of Systems', *Strategic Forum*, no. 63, February 1996.

19 The three books are published by the Command and Control Research Project managed by Evidence Based Research (EBR). While EBR is an independent think tank, the presence of Dr David Alberts speaks to the authority of these works. At the time, Alberts was Director of Research and Strategic Planning in the Office of the Assistant Secretary of Defense (C3I). David S. Alberts, John J. Gartska, and Frederick P. Stein, *Network Centric Warfare: Developing and Leveraging Information Superiority*, 2nd edition, Washington DC: Command and Control Research Program, 1999; David S. Alberts, John J. Gartska, Richard E. Hayes, and David A. Signori, *Understanding Information Age Warfare*, Washington DC: Command and Control Research Program, 2001; David S. Alberts and Richard E. Hayes, *Power to the Edge: Command and Control in the Information Age*, Washington DC: Command and Control Research Program, 2003.

20 Alvin and Heidi Toffler, *War and Anti-war: Survival at the Dawn of the 21st Century*, Boston, MA: Little, Brown, 1993, p. 80.

21 Alberts *et al.*, *Network Centric Warfare*, p. 54.

22 Ibid., pp. 60–65.

23 Ibid., pp. 71–72.

24 Ibid., p. 41.
25 Ibid., p. 90.
26 Alberts *et al.*, *Understanding Information Age Warfare*, pp. 14–21.
27 Ibid., pp. 12–13.
28 Ibid., pp. 15–18.
29 Department of Defense, *Network Centric Warfare Report to Congress*, July 2001.
30 Alberts *et al.*, *Understanding Information Age Warfare*, p. 26.
31 Ibid., p. 60.
32 Colonel George K. Gramer (USA), 'Optimizing Intelligence Sharing in a Coalition Environment: Why US Operational Commanders Have an Intelligence Dissemination Problem', course paper, Department of Joint Military Operations, US Naval War College, Newport, RI, 17 May 1999, pp. 2–3.
33 Alberts *et al.*, *Understanding Information Age Warfare*, pp. 57–58.
34 Ibid., pp. 12–13. Emphasis added.
35 Alberts and Hayes, *Power to the Edge*, p. 56.
36 Ibid., p. 59.
37 Ibid., pp. 4–5.
38 Ibid., p. 187.
39 Paul Wolfowitz, 'Global Information Grid (GIG) Overarching Policy', *Department of Defense Directive 8100.1*, 19 September 2002, www.dtic.mil/whs/directives/corres/html2/d81001x.htm.
40 Committee on Network-Centric Naval Forces, Naval Studies Board, *Network Centric Naval Forces: A Transition Strategy for Enhancing Operational Capabilities*, Washington DC: National Academy Press, 2000, p. 31.
41 Statement by John P. Stenbit before the Committee on Armed Services, United States House of Representatives, Terrorism, Unconventional Threats and Capabilities Subcommittee, 11 February 2004.
42 Robert E. Levin, *The Global Information Grid and Challenges Facing Its Implementation*, GAO 84–858, Washington DC: Government Accounting Office, July 2004, p. 1.
43 The GIG-BE is a worldwide ground-based fiber-optic network, using IP protocols, to expand the connectivity and interoperability of DOD installations. Six sites achieved initial operating capability on 30 September 2004. 'Global Information Grid (GIG) Bandwidth Expansion (GIG-BE)', www.globalsecurity.org/space/systems/gig-be.htm. See also Statement by John P. Stenbit.
44 The TCS comprises space-based and ground-based segments. Space-based segments include the Transformation Satellite (TSAT) and Advanced Polar System (APS) satellites, a laser-based SATCOM constellation allowing global IP routing and addressing of information, even in areas with no pre-existing communications infrastructure. The ground-based segment comprises the Joint Tactical Radio System (JTRS), a software-based radio that will be programmable to imitate other types of radios thus enhancing overall communications interoperability within the US military. Able to transmit voice, data, and video, it is hoped that JTRS will enable seamless communication, hypothetically between fighter pilots to soldiers to sailors. See Jefferson Morris and Rich Tuttle, 'Contractors Lining Up To Compete for Transformational Communications Network', *Aerospace Daily*, vol. 207, no. 38, p. 1; Robert E. Levin, *The Global Information Grid*, pp. 11–12; Johnny Kegler, 'Pathways to Enlightenment', *Armada International*, vol. 29, no. 5, October–November 2005, pp. 10–14; Johnathon Karp and Andy Pasztor, 'Pentagon Week: High Tech Has High Risk', *Wall Street Journal*, 2 May 2005, p. B2; 'Transformational Communications Architecture', www.globalsecurity.org/space/systems/tca.htm; 'Transformational SATCOM (TSAT) Advanced Wideband System', www.globalsecurity.org/space/systems/tsat.htm.
45 NCES are the integrated series of applications that will reside on the GIG permitting the military to access, send, store, and protect information. In effect, this will create the software 'nervous system' that will operate the GIG. By establishing IP protocols

on the GIG, NCES will enable US forces to forego the typical 'point to point' interfaces between systems, ending duplication of effort and the multiplication of incompatible systems. Levin, *The Global Information Grid*, p. 11; 'Global Information Grid (GIG)', www.globalsecurity.org/space/systems/gig.htm.

46 *Network Centric Naval Forces*, p. 3.

47 Committee to Review DOD C4I Plans and Programs, Computer Science and Telecommunications Board, National Research Council, *Realizing the Potential of C4I*, Washington DC: National Research Council, 1999, p. 70.

48 Ibid., p. 27, Alberts *et al.*, *Network Centric Warfare*, pp. 60–65.

49 Cohen, 'Change and Transformation in Military Affairs', p. 395; Alberts and Hayes, *Power to the Edge*, p. 88.

50 General Charles C. Krulak (USMC), 'The Strategic Corporal: Leadership in the Three Block War', *Marines Magazine*, January 1999.

51 Alberts *et al.*, *Network Centric Warfare*, pp. 20–21; *Network Centric Naval Forces*, p. 3.

52 Manuel Castells, 'Informationalism, Networks and the Network Society: A Theoretical Blueprint', in Manuel Castells (ed.), *The Network Society: A Cross-Cultural Perspective*, Cheltenham: Edgar Elgar, 2004, pp. 3, 5–6.

53 Statement by John P. Stenbit.

54 'Global Information Grid (GIG)'. Alberts and Hayes point out in *Power to the Edge* that expanding access to information eliminates 'unnecessary constraints previously needed to deconflict elements of the force in the absence of quality information' (p. 5).

55 *Transformation Planning Guidance*, p. 3; Cohen, 'Change and Transformation in Military Affairs', p. 1.

56 Alberts *et al.*, *Network Centric Warfare*, p. 71.

57 'A process that shapes the changing nature of military competition and cooperation through new combinations and concepts, capabilities, people, and organisations that exploit our nation's advantages, protect against our asymmetric vulnerabilities to sustain our strategic position which helps underpin peace and stability in the world.' *Transformation Planning Guidance*, p. 3.

58 Levin, *The Global Information Grid*, p. 1.

59 Alberts *et al.*, *Network Centric Warfare*, p. 54.

60 *Network Centric Naval Forces*, p. 59.

61 Charlotte Adams, 'Network Centric Rush To Connect', *Aviation Today*, 1 September 2004.

62 *Realizing the Potential of C4I*, p. 135.

63 Ibid., p. 143. See also Duane P. Andrews (chairman), *Report of the Defense Science Board Task Force on Information Warfare Defense*, Washington DC: Defense Science Board, November 1996, pp. 37–45, http://cryptome.org/iwdmain.htm.

64 One is tempted to argue against the possibility of establishing a digital identity. Human beings are essentially analogue entities – unique and discrete. Digital entities, through their ordinal precision and endlessly replicable nature, mean such a fundamental identification will prove elusive in its very essence.

65 Major Joshua Reitz (USA), *Untangling the Web: Balancing Security, Prosperity, and Freedom in the Information Age*, MDS dissertation, Toronto: Canadian Forces College, May 2005, pp. 11–14.

66 According to the GAO, draft readiness metrics went untested, and organisational policies and procedures for managing information assurance were not fully defined across the DOD. See Robert F. Dacey, *Progress and Challenges to an Effective Defense-wide Information Assurance Program*, GAO-01–307, Washington DC: GAO, March 2001, p. 4.

67 Levin, *The Global Information Grid*, p. 19.

68 Adams, 'Network Centric Rush to Connect'. Reportedly, JTRS radios would be able to 'firewall' information within transmissions. In this way, information would be double-encrypted in terms of both data and transmission.

69 Wolfowitz, 'Information Assurance', *Department of Defense Directive 8500.1*, 24 October 2002, p. 20.
70 Dacey, *Progress and Challenges*, p. 6.
71 As one study examining the impact of networks on naval forces argues: 'Strict controls will be necessary at the connection points between tactical and non-tactical portions of the Naval Command and Information Infrastructure. These controls will ensure that only authorised types of traffic are allowed onto the tactical networks, and hence they will provide continued guarantees that the tactical networks can provide highly reliable, low latency data services. These controls will also aid in providing security boundaries'. *Network Centric Naval Forces*, p. 33.
72 Levin, *The Global Information Grid*, pp. 28–29.
73 Joe Pappalardo, 'Protecting GIG Requires a New Strategy', *National Defence*, October 2005.
74 Hedley Bull, *The Anarchical Society, A Study of Order in World Politics*, London: Macmillan, 1977.
75 This is not strictly true in some parts of Asia, where the state has retained a degree of control over Internet communications.
76 Robert Burnett and P. David Marshall, *Web Theory: An Introduction*, London: Routledge, 2003, pp. 32–33; 'Wikipedia Study "Fatally Flawed"', *BBC News*, http://news.bbc.co.uk/2/hi/technology/4840340.stm.
77 Brock Read, 'Can Wikipedia Ever Make the Grade?', *Chronicle of Higher Education*, 27 October 2006, http://chronicle.com/tem/reprint.php?%20id=z6xht2rj60kqmsl8tlq5ltqcshc5y93y; see also 'Internet Encyclopaedias Go Head to Head', *Nature*, 15 December 2005, www.nature.com/nature/journal/v438/n7070/full/438900a.html.
78 For example, www.911truth.org/, www.aliensthetruth.com/ and www.anomalies.net/area51/faq.
79 As Morgenthau puts it:

> Where the insecurity of human existence challenges the wisdom of man, there is the meeting point of fate and freedom, of necessity and chance. Here, then, is the battlefield where man takes up the challenge and joins battle with the forces of nature, his fellow-men's lust for power, and the corruption of his own soul.
>
> (Hans Morgenthau, *Scientific Man vs. Power Politics*, Chicago, IL: University of Chicago Press, 1946, p. 223)

See also E. H. Carr, *The Twenty Years Crisis 1919–1939*, New York: Harper and Row, 1964, pp. 63–88; Michael Howard, 'Morality and Force in International Politics', *Studies in War and Peace*, London: Temple Smith, 1970, pp. 235–250.
80 Castells, 'Informationalism, Networks and the Network Society', pp. 17–21.
81 Himanen uses the term hacker ethic, although he notes that the negative connotations that come with the term 'hacker' have distorted its original meaning as an informal society of technologically savvy and creative individuals intent on the propagation of truth through the free sharing of information. See Pekka Himanen, 'The Hacker Ethic as the Culture of the Information Age', in Castells (ed.), *The Network Society*, p. 424.
82 Ibid., p. 423.
83 See, for example, www.opensource.org and www.linux.org/lininfo/index.html.
84 See Vincent Moscoe, *The Digital Sublime: Myth, Power, and Cyberspace*, Cambridge, MA: MIT Press, 2004.
85 See www.opensource.org/advocacy/secrets.php for example. The limitation on secrecy and knowledge is reached when many similar products are circulating performing similar services; at that point, they argue, it makes more sense to open up research in order that products and services can be improved through information sharing.
86 As described by Richard Hunter, Open Source development is guided by 'extraordinary talent, clear vision of the goal, a deadly enemy, extraordinary tools, and

autonomy and responsibility'. Richard Hunter, *World Without Secrets: Business, Crime, and Privacy in the Age of Ubiquitous Computing*, New York: John Wiley and Sons, 2002, p. 97. *Auftragstaktik's* decentralised approach to operations devolves a significant amount of creative freedom all down the command hierarchy even into the ranks of non-commissioned officers. Robert Leonhard, *The Art of Maneuver: Maneuver Warfare Theory and AirLand Battle*, Novato Ca: Presidio Press, 1991, pp. 50–51.

87 Burnett and Marshall, *Web Theory*, pp. 27–28. A classic example of this problem is the misinterpretation of sensor data by CIC operators aboard the *USS Vincennes* in 1988 during operations in the Persian Gulf, when a civilian airliner was portrayed by the system as an F-14 fighter jet. See Marita Turpin and Niek du Plooy, 'Decision-Making Biases and Information Systems', *Decision Support in an Uncertain and Complex World: The IFIP TC8/WG8.3 International Conference*, http://vishnu.sims.monash.edu.au:16080/dss2004/proceedings/pdf/77_Turpin_Plooy.pdf.

88 See, for example, Sayaka Kawakami and Sarah C. McCartney, 'Government Information Collection: Privacy Year in Review: Privacy Impact Assessments, Airline Passenger Pre-Screening, and Government Data Mining', *I/S: A Journal of Law and Policy for the Information Society*, vol. 1, nos 2–3, Spring/Summer 2005, pp. 245–256; Michael J. Sniffen, 'Controversial Government Data Mining Research Lives On', 23 February 2004, www.kdnuggets.com/news/2004/n05/20i.html; Max Blumenthal, 'Data Debase', *American Prospect*, 19 December 2003, www.prospect.org/webfeatures/2003/12/blumenthal-m-12–19.html.

89 Castells, 'Informationalism, Networks and the Network Society', p. 12.

90 Peter Howard, 'The USN's Designer of Concepts', *Jane's Defence Weekly*, 3 October 2001.

91 Castells 'Informationalism, Networks and the Network Society', p. 23.

92 Ibid., p. 29.

93 See, for example, David A. Powner and Eileen Laurence, *Information Sharing: The Federal Government Needs to Establish Policies and Processes for Sharing Terrorism-Related and Sensitive but Unclassified Information*, GAO-06–385, Washington DC: GAO, March 2006.

3 International anarchy and military cooperation

1 Frederick Kagan, 'The Military's Manpower Crisis', *Foreign Affairs*, vol. 85, no. 4, July–August 2006.

2 Australia, Great Britain, Canada, the United States, France, Germany, and Italy.

3 The exception may be Sweden, which is implementing a mature network-centric system in its own armed forces. Still, Sweden has a long tradition of neutrality and coalition operations do not inform its operational ethos in the way that they do the operations of major NATO partners.

4 Glenn H. Snyder, *Alliance Politics*, Ithaca, NY: Cornell University Press, 1997, p. 17.

5 Ibid., p. 17.

6 Charles W. Kegley Jr. and Gregory A. Raymond, *When Trust Breaks Down: Alliance Norms and World Politics*, Columbia, SC: University of South Carolina, 1990, p. 55.

7 Snyder, *Alliance Politics*, p. 170.

8 Stephen Walt, *The Origins of Alliances*, Ithaca, NY: Cornell University Press, 1987, p. 43; Snyder, *Alliance Politics*, p. 171.

9 Kegley and Raymond, *When Trust Breaks Down*, pp. 266–267.

10 Steven Metz, 'The Effect of Technological Asymmetry on Coalition Operations', in Thomas J. Marshall, Phillip Kaiser, and Jon Kessmeier (eds), *Problems and Solutions in Future Coalition Operations*, Carlisle, PA: US Army War College Strategic Studies Institute, December 1997, p. 56.

11 Kenneth Gause, 'US Navy Interoperability with Its High-End Allies', unpublished paper, p. 7.

12 Peacetime alliances generally limit themselves to defensive pacts calling for mutual support in case of attack, non-aggression treaties, or limited ententes. Kegley and Raymond, *When Trust Breaks Down*, p. 53.
13 Hans. J. Morgenthau, 'Alliances', in Julian R. Friedman, Christopher Bladen, and Steven Rosen (eds), *Alliance in International Politics*, Boston, MA: Allyn and Bacon Inc., 1970, p. 84.
14 John Garnett, 'Limited War', in John Baylis, Ken Booth, John Garnett, and Phil Williams (eds), *Contemporary Strategy: Theories and Policies*, Beckenham: Croom Helm, 1975, pp. 122–124.
15 Carl von Clausewitz, *On War*, Princeton, NJ: Princeton University Press, 1976, p. 603.
16 Ibid., p. 596.
17 Robert Osgood, *Alliances and American Foreign Policy*, Baltimore, MD: Johns Hopkins University Press, 1968, p. 5.
18 See especially, Ivo Daalder and Michael O'Hanlon, *Winning Ugly*, Washington DC: Brookings, 2001.
19 Nineteen old Joint Operational Tactical System terminals were given to NATO command centres to support the maritime interdiction operation against the former Yugoslavia *Operation Sharp Guard*, for example. See Eric Francis Germain, 'The Coming Revolution in NATO Maritime Command and Control', *Mitre Technical Papers*, www.mitre.org/support/papers/technet97/germain_technet.pdf.
20 Paul T. Mitchell, 'Small Navies and NCW: Is There a Role?', *Naval War College Review*, vol. 61, no. 2, Spring 2003.
21 See Gary McKerow, 'Multilevel Security Networks: An Explanation of the Problem', *SANS Information Security Reading Room*, 5 February 2001, www.sans.org/reading_room/whitepapers/standards/546.php, p. 2; S. C. Spring *et al.*, 'Information Sharing for Dynamic Coalitions', unpublished paper, Pacific Sierra Research, Arlington, VA, December 2000, pp. 29–34; Colonel Robert Chekan, 'The Future Of Warfare: Clueless Coalitions?', course paper, Canadian Forces College, October 2001, pp. 9–23.
22 Chekan, 'The Future Of Warfare', p. 11.
23 Henry S. Kenyon, 'Alliance Forces Move Toward Unified Data Infrastructure', *Signal*, vol. 56, no. 1, September 2001, p. 21.
24 Quoted in Lieutenant-Commander Thomas Spierto, 'Compromising the Principles of War: Technological Advancements Impact Multinational Military Operations', course paper, Naval War College, Newport, RI, 5 February 1999, p. 3.
25 See, for example, Robert W. Riscassi, 'Principles for Coalition Warfare', *Joint Forces Quarterly*, no. 1, Summer 1993.
26 Chekan, 'The Future Of Warfare', p. 4.
27 Lieutenant-Colonel William R. Pope, 'US and Coalition Command and Control Interoperability for the Future', thesis, US Army War College, Carlisle, PA, April 2001, p. 6.
28 'General Warns over Digitization Split', *International Defence Review*, 1 January 2002; John Kiszely, 'Achieving High Tempo: New Challenges', *RUSI Journal*, vol. 144, no. 6, December 1999.
29 Commander James Carr, 'Network Centric Coalitions: Pull, Pass, or Plug-in?', course paper, Naval War College, Newport, RI, 15 May 1999, pp. 15–16.

4 Naval networks in the coalition environment

1 Geraghty notes that a cautious co-existence between NCW and coalition operations might ultimately evolve, much like the issue of coalition command authorities that persistently bedevils multinational military operations. However, many are not so sanguine. Pope argues that the potential for failure in these types of operations is growing. Carr describes a 'gaping mismatch' between the demands of operational doctrine and the strategy of operating in coalitions. This mismatch is driving a 'widening interoperability chasm' threatening America's ability to operate within

coalitions. Commander Barbara A. Geraghty (USN), 'Will Network Centric Warfare be the Death Knell for Allied/Coalition Operations?', course paper, Department of Joint Military Operations, US Naval War College, Newport, RI, 17 May 1999, p. 15; Lieutenant Colonel William R. Pope, (USA), 'US and Coalition Command and Control Interoperability for the Future', thesis, U.S. Army War College, Carlisle, Pa., April 2001, p. 19; Carr, 'Network Centric Coalitions', p. 19.

2 Captain Robert M. Stuart (USN), 'Network Centric Warfare in Operation Allied Force: Future Promise or Future Peril?', course paper, Department of Joint Military Operations, US Naval War College, Newport, RI, 16 May 2000, p. 15.

3 Major Michael B. Black (USA), 'Coalition Command, Control, Communications, Computer and Intelligence Systems Interoperability: A Necessity or Wishful Thinking?', thesis, US Army Command and General Staff College, Fort Leavenworth, KS, 2 June 2000, p. 66.

4 Major Robert L. Coloumbe (USMC), 'Operational Art and NATO C4I: An Oxymoron?', course paper, Department of Joint Military Operations, US Naval War College, Newport, RI, 5 February 2001, pp. 17–18.

5 Commander J. L. R. Foreman (RN), 'Multinational Information Sharing (MNIS)', unpublished briefing slides, pp. 3–4.

6 Each of these terms are 'classification markings' that are 'a documentary form of classification guidance issued by an original classification authority that identifies the elements of information regarding a specific subject that must be classified and establishes the level and duration of classification for each such element.' See: Susan Maret, *On Their Own Terms: A Lexicon with an Emphasis on Information-Related Terms Produced by the U.S. Federal Government*, www.fas.org/sgp/library/ maret.pdf, pp. 43–64.

7 For example, DCID1/7 suggests that material be produced in a 'collateral uncaveated level to the greatest extent possible without diluting the meaning of the intelligence'. Where this is not possible, intelligence reports should use 'tear lines' that identified those items that could not be shared and those that could. See Director of Central Intelligence Directive 1/7, 'Security Controls on the Dissemination of Intelligence Information', 15 June 1996, Sections 7 and 12, www.fas.org/irp/offdocs/ dcid17m.htm.

8 'General Warns over Digitisation Split'; Kiszely, 'Achieving High Tempo'; Smith, 'Network-Centric Warfare', p. 3; Oxendine, 'Managing Knowledge', p. 19.

9 It is important to note, however, that where there is a 'need to know', the US will provide limited access to raw SIGINT data. Mark MacIntyre and Sherri Flemming, 'Netcentric Warfare for Dynamic Coalitions: Implications for Secure Interoperability', paper presented at the RTO IST Symposium on 'Information Management Challenges in Achieving Coalition Interoperability', Québec, 28–30 May 2001, pp. 21–24.

10 Susan C. McGovern, *Information Security Requirements for a Coalition Wide Area Network*, Masters thesis, Naval Post-Graduate School, Montery, June 2001, p. 38. McGovern goes on to note that, given this constraint, information is released at the 'highest level of clearance common to all members'.

11 At the time of publication, there are reports that the US is permitting access to the SIPRNET to its Australian, British, and Canadian partners. It is not clear how extensive that access is. Further, it appears that such access has not been extended to other American coalition partners. See: David E. Kaplan and Kevin Whitelaw, 'Remaking US Intelligence', *US News and World Report*, 3 November 2006, www.usnews.com/usnews/news/articles/061103/3dni.intro.htm.

12 'The Combined Communications and Electronics Board (CCEB) is a five nation joint military communications-electronics (C-E) organisation whose mission is the coordination of any military C-E matter that is referred to it by a member nation. The member nations of the CCEB are Australia, Canada, New Zealand, the United Kingdom and the United States of America. The CCEB Board consists of a senior Command, Control, Communications and Computer (C4) representative from each of the member nations.' 'The Multinational Interoperability Council (MIC) is a

multinational, operator-led forum, to identify interoperability issues and articulate actions, which if nationally implemented, would contribute to more effective coalition operations. While initial work focused on resolving information interoperability problems, the scope of the MIC has expanded to cover other strategic and operational issues considered key to coalition operations. The MIC member nations are Australia, Canada, France, Germany, Italy, the United Kingdom, and the United States which are nations most likely to form, lead and/or support coalition operations. New Zealand and NATO Allied Command Transformation (ACT) have official observer status in the MIC. The MIC is composed of senior operations, doctrine, logistics, and C4 staff officers from each of the member nations as well as senior officials from observer nations and organizations.' See www.jcs.mil/j6/cceb/ and www.jcs.mil/j3/mic. A number of other organisations are devoted to the problems of allied and coalition interoperability, such as ABCA and AUSCANUKUS; there are links to these bodies from these web pages.

13 CCEB, *A Strategy for Improved Coalition Networking*, June 2005, p. 1, www.jcs.mil/j6/cceb/cnsdatedjune05.pdf.

14 McGovern, *Information Security Requirements*, p. 21.

15 CCEB, *A Strategy for Improved Coalition Networking*, pp. 1–2.

16 MIC, *Report of the Multinational Interoperability Council, 27–28 October 1999*, 1 March 2000, p. 10.

17 MIC, *Report on MIC 2000, November 8–9, 2000*, 19 January 2001, p. 10.

18 MIC, *Report on MIC 2002, April 16–18, 2002*, 7 June 2002, p. 9.

19 CCEB, *A Strategy for Improved Coalition Networking*, pp. 2–3.

20 The MIC authorised the establishment of the first COWAN in October 1999 in its efforts to improve collaborative planning activities. The MIC noted: 'The C[O]WAN when fully implemented, will provide an apparently seamless and robust network capability of exchanging and sharing information that is operationally relevant to all coalition partners involved in multinational operations.' Ibid., p. 4.

21 Briefing note for Lieutenant-Colonel B. Green, (CF), ABCA, undated, p. 1.

22 Thomas MacIntyre, 'CENTRIXS Improves Communication for RIMPAC 2004', www.news.navy.mil, Story Number NNS040707–28, 8 July 2004.

23 *Griffin Key Attributes*, 25 January 2005, www.jcs.mil/j6/cceb/griffinkeyattributes 26jan05.pdf.

24 The Australian Navy maintained frigates in the Persian Gulf and Red Sea throughout the 1990s, supporting the Maritime Interdiction Force enforcing various UN Security Council Resolutions under the rubric of *Operation Damask*. Canada also sent frigates for similar purposes throughout the 1990s under a variety of different operation code names. Starting in 1995, Canadian frigates began to be integrated into US carrier battle groups. Greg Nash and David Stevens, *Australia's Navy in the Gulf*, Silverwater: Topmill, 2006, pp. 36–43; Richard Gimblett, *Operation Apollo*, Ottawa: Magic Light, 2004, pp. 32–37; Mitchell, 'Small Navies and NCW'.

25 Nash and Stevens, *Australia's Navy in the Gulf*, pp. 36–43.

26 Commodore Eric Lerhe (CF) and CPO2 Doug McLeod (CF), 'Canadian Naval Task Groups in Op Apollo', *Maritime Tactical Warfare Bulletin*, 2003, p. 1.

27 James Goldrick notes: 'The battlespace was measured in just a few miles and the time available was minutes rather than hours. We could not afford mistakes.' James Goldrick, 'In Command in the Gulf', *US Naval Institute Proceedings*, vol. 128, no. 12, December 2002. Interview with Rear-Admiral James Goldrick (RAN), Canberra, 30 May 2006.

28 Commander John Bycroft (CF), 'Coalition C4ISTAR Capability AUSCANUKUS', unpublished paper presented to the SMi conference 'Naval C4ISTAR', London, 21 April 2004, p. 4.

29 Rear-Admiral Thomas E. Zelibor (USN), 'FORCEnet is Navy's Future: Information Sharing from Seabed to Space', *Armed Forces Journal*, December 2003, www.chinfo.navy.mil/navpalib/.www.rhumblines/rhumblines170.doc.

30 Captain Paul Maddison (CF), 'The Canadian Navy's Drive for Trust and Technology in Network Centric Coalitions: Riding Comfortably Alongside, or Losing Ground in a Stern Chase?', course paper, Canadian Forces College, 2004, p. 17. Lerhe noted to the author: 'In a large measure I believe his [Zelibor's] view is that of an East coast ship that continued to lag the West coast fleets NCW progress. I suspect his ship was thrown in at the last minute into a confusing *Operation Iraqi Freedom* picture where the USN was necessarily rebuilding its networks. Moreover, they were concentrating on Iraq and thus the UK and Australia. During my watch COWAN was where the real battle during *Operation Enduring Freedom* was fought and there is no doubt whatsoever about that ... my situational awareness was likely better than the USN's in this most critical of contact sets.' E-mail from Commodore Eric Lerhe (CF Ret.d) to the author, 10 August 2006.

31 'Despite the CFLCC C-5 Planner's best efforts, he could not get through the restrictive administration required to become registered as a SIPRNET CENTRIXS – X user.' Lieutenant-Colonel Chris Field (ADF), 'An Australian Defence Force Liaison Officer's Observations and Insights from Operation Iraqi Freedom', *Australian Defence Force Journal*, no. 163, November–December 2003, p. 5.

32 Interview with Commodore Peter Jones (RAN), Canberra, 2 June 2006; interview with Commodore Eric Lerhe (CF Ret.d), Halifax, NS, 30 September 2005.

33 Interview at the Australian Air Power Development Centre, Tuggeranong, Australia, 31 May 2006; interview with Rear-Admiral Drew Robertson (CF), Ottawa, 28 September 2005.

34 Interview with Rear-Admiral James Goldrick (RAN), 30 May 2006.

35 Interview with Major-General Angus Watt (CF), Ottawa, 28 September 2005.

36 Bycroft, 'Coalition C4ISTAR Capability', p. 4; interview with Major-General Angus Watt (CF), Ottawa Canada, 28 September 2005; interview with Lieutenant Commander Mark DeSmedt (CF), Ottawa, 28 September 2005.

37 Interview with Major-General Angus Watt (CF), 28 September 2005.

38 Interview with Air Commodore Mark Lax (RAAF), Canberra, 31 May 2006; interview with Major-General Angus Watt (CF), 28 September 2005.

39 Interview with Captain Phillip Spedding (RAN), Canberra, 1 June 2006.

40 Interview with Major-General Angus Watt (CF), 28 September 2005.

41 Interview with Captain Phillip Spedding (RAN), 1 June 2006.

42 Ibid.

43 Allan English, Richard Gimblett and Howard Coombs, *Beware of Putting the Cart Before the Horse: Network Enabled Operations as a Canadian Approach to Transformation*, DRDC Contract Report CR 2005–212, Toronto, 19 July 2005, p. 13, http://pubs.drdc-rddc.gc.ca/pubdocs/pcow1_e.html.

44 Richard Gimblett, 'Command of Coalition Operations in a Multicultural Environment: A Canadian Naval Niche? The Case Study of Operation Apollo', unpublished paper prepared for the Canadian Forces Leadership Institute.

45 Multiplexing a satellite channel allows several different communication streams to be run on the same channel. Thus, a multiplexed satellite channel might have 70 per cent of its capacity devoted to a national secret level network, and the remaining 30 per cent devoted to a national unclassified administrative network.

46 Interview with Commodore Eric Lerhe (CF Ret.d), 30 September 2005.

47 Lieutenant Michael Parker (RAN), 'RAN Exercises', *Journal of the Australian Naval Institute*, no. 115, Summer 2005, p. 30.

48 Interview with Rear-Admiral James Goldrick (RAN), 30 May 2006.

49 The number of satellite channels is dependent on the capacity of communication satellites already in geo-stationary orbit, a resource that cannot be expanded rapidly. Interview with Rear-Admiral James Goldrick (RAN), 30 May 2006; interview with Commodore Peter Jones (RAN), 2 June 2006.

50 These included CENTRIXS, CENTRIXS GFE, CENTRIXS J, CENTRIXS C, and CENTRIXS R. English, Gimblett, and Coombs, *Beware of Putting the Cart Before the Horse*, p. 15.
51 Interview with Rear-Admiral Drew Robertson (CF), 28 September 2005.
52 Ibid.
53 Multilevel security would allow sharing of information on networks between individuals, organisations, and nations, all cleared for differing levels of classification. Interview with Commodore Eric Lerhe (CF Ret.d), 30 September 2005.
54 Interview with Rear-Admiral Drew Robertson (CF), 28 September 2005; interview with Commodore Peter Jones (RAN), 2 June 2006; Lieutenant-Commander Ivan Ingham (RAN), 'Naval Gunfire Support for the Assault of the Al Faw Peninsular', *Journal of the Australian Naval Institute*, no. 109, Winter 2003, p. 36.
55 Interview with Commodore Eric Lerhe (CF Ret.d), 30 September 2005.
56 Interview with Rear-Admiral Drew Robertson (CF), 28 September 2005.
57 Interview with Commodore Eric Lerhe (CF Ret.d), 30 September 2005.
58 Interview with Rear-Admiral Drew Robertson (CF), 28 September 2005; interview with Commodore Eric Lerhe (CF Ret.d), 30 September 2005.
59 Ingham, 'Naval Gunfire Support for the Assault of the Al Faw Peninsular', p. 34.
60 Interview with Commodore Eric Lerhe (CF Ret.d), 30 September 2005.
61 Captain Jones sent his own chief of staff, somebody 'ugly enough and strong enough to give honest answers to an Adm. and come back and tell me what I was doing was wrong'. Commodore Lerhe noted that 'if it doesn't hurt [in terms of human resources] to send liaison officers, then you are sending either the wrong people, or not enough of them'. Interview with Commodore Peter Jones (RAN), 2 June 2006; interview with Commodore Eric Lerhe (CF Ret.d), 30 September 2005.
62 Field, 'An Australian Defence Force Liaison Officer's Observations', p. 11.
63 Interview at the Air Power Development Centre, Tuggeranong Australia, 31 May 2006; interview with Captain Phillip Spedding (RAN), 1 June 2006.
64 Ibid.; Field, p. 11.
65 Alan Ryan, 'Australian Army Cooperation with the Land Forces of the United States: Problems of a Junior Partner', *Land Warfare Studies Centre Working Paper*, no. 121, January 2003, p. 4.
66 Gimblett, *Operation Apollo*, p. 108.
67 Gimblett, 'Command of Coalition Operations in a Multicultural Environment', p. 13.
68 Interview at the Air Power Development Centre, Tuggeranong Australia, 31 May 2006. Commodore Steve Gilmore noted that, in the planning of coalition operations, knowledge of a nation's ROE was as important as understanding the capabilities of the type of kit and the professionalism of the crews they sent. Interview Commodore Steve Gilmore (RAN), Canberra, 2 June 2006.
69 Captain Phil Wisecup and Lieutenant Tom Williams (USN), 'Enduring Freedom: Making Coalition Naval Warfare Work', *Proceedings*, vol. 128, no. 9, September 2002, p. 55.
70 Commodore Eric Lerhe (CF Ret.d), 'Multilateralism and Interoperability: Impact on Maritime Capabilities', paper presented to the Centre for Foreign Policy Studies conference 'What Canadian Military and Security Forces in the Future World? A Maritime Perspective', Halifax, NS, 10–12 June 2005, pp. 8–9, http://centreforforeignpolicystudies.dal.ca/pdf/msc2005/msc2005lerhe.pdf.
71 Interview with Commodore Eric Lerhe (CF Ret.d), 30 September 2006.
72 Ibid.
73 Interview with Rear-Admiral Drew Robertson (CF), 28 September 2005. Similar issues were raised by James Goldrick, Peter Jones, and Steve Gilmore in their interviews.
74 Interview with Commodore Eric Lerhe (CF Ret.d), 30 September 2005.
75 English, Gimblett, and Coombs, *Beware of Putting the Cart Before the Horse*, p. 14.
76 Interview with Commodore Peter Jones (RAN), 2 June 2006.

77 Interview with Commodore Steve Gilmore (RAN), 2 June 2006; interview with Commodore Peter Jones (RAN), 2 June 2006; Zelibor, 'FORCEnet is Navy's Future'.

5 The neighbourhood watch: organisational and political boundaries in NORAD

1 The difference between Air Operations Centres and Combined Air Operations Centres is simply the addition of coalition or allied partners within the structure of the organisation. AOC and CAOC will be used interchangeably, therefore.
2 David Fulghum, 'A Crowded Room', *Aviation Week and Space Technology*, vol. 60, no. 17, 26 April 2006, p. 52.
3 Allied personnel are typically assigned as liaison to assist with the integration of their national efforts into the overall air operation. Those forces contributing critical numbers or assets may be assigned positions of greater importance within the CAOC. However, a review of operations in both 1991 and 2003 illustrates the overwhelming nature of the American effort as contrasted with allied efforts. For example, in 1991, coalition partner operations accounted for 20 per cent of the total number of sorties. In 2003, the RAF and RAAF accounted for only 10 per cent of the total number of sorties executed by 25 April 2003. In OIF, of 9,920 tanking sorties, the RAF only flew four; of the 12,500 transport sorties, the USAF flew all but 750 of them. Bruce Rolfsen, 'Air Power Unleashed – Lessons from Iraq', *Armed Forces Journal*, 1 June 2003.
4 LCol. Joseph H. Justice III (USAF), *Air Power Command and Control: Evolution of the Air and Space Operations Center as a Weapon System*, US Army War College Research Project, Carlisle Barracks PA: US Army War College, 3 May 2004, p. 2.
5 Winnefeld and Johnson provide a detailed overview of the many doctrinal and organisational issues which have complicated efforts to bring unity in command and control to air operations. See, James A. Winnefeld and Dana J. Johnson, *Joint Air Operations: Pursuit of Unity in Command and Control*, Annapolis: Naval Institute Press, 1993.
6 Ibid., p. 1.
7 Marvin Simpson and Leonard Simpson, 'Bettering National Response by Effectively Using the CAOC', paper presented at the Command and Control Research and Technology Symposium, 2006, p. 2.
8 David Fulghum, 'USAF Streamlines the Air Operations Center', *Aviation Week and Space Technology*, vol. 157, no. 13, 23 September 2002, pp. 53–56.
9 Simpson and Simpson, 2006, p. 3.
10 Gordon Trowbridge, 'Bringing Order from Chaos', *Air Force Times*, 20 December 2004.
11 Ibid.
12 Justice, 2004, pp. 5–6.
13 Mark Hewish, 'Out of CAOCs Comes Order', *International Defence Review*, 1 May 2003.
14 Simpson and Simpson, 2006, p. 3.
15 'Theatre Battle Management Core Systems', www.globalsecurity.org/military/systems/aircraft/systems/tbmcs.htm.
16 'Automated Deep Operations Coordination System', www.globalsecurity.org/military/systems/ground/adocs.htm.
17 Joris Janssen Lok, 'Communication Weaknesses Endanger Allied Integration in US led Air Campaigns', *International Defence Review*, 1 March 2004.
18 Joris Janssen Lok, 'Next Level Needed for NATO ACCS', *International Defence Review*, 1 July 2002.
19 Lok, 1 March 2004.
20 Robert Akerman, 'Aerospace Experts Refocus the Tactical Picture', *Signal*, vol. 55, no. 3, November 2000, pp. 23–27.
21 Lorenzo Cortes, 'CAOC Crews Credit TBMCS and IWS for OIF Success', *C4I News*, 12 December 2003, p. 1.

22 Rolfsen, 'Air Power Unleashed – Lessons from Iraq', 2003.
23 Lok, 1 March 2004.
24 LCdr. Jonathon Lee Jackson (USN), *Solving the Problem of Time Sensitive Targets*, Joint Military Operations Paper, US Naval War College, 3 February 2003, pp. 2–6.
25 David Fulghum, 'New Bag of Tricks', *Aviation Week and Space Technology*, vol. 158, no. 16, pp. 22–25.
26 Rolfsen, 'Air Power Unleashed – Lessons from Iraq', 2003.
27 Jackson, 2003, p. 2.
28 Hewish, 1 May 2003.
29 David Fulghum, 'A Crowded Room', 2006, p. 54.
30 Capt. Marcella Adams (USAF), 'Controlling the Bosnian Skies', *Airman*, available at www.af.mil/news/airman/0896/caoc.htm.
31 Mustafa R. Koprucu, *The Elements of Decentralized Execution: the Effect of Technology on a Central Air Power Tenet*, thesis prepared for the School of Advanced Airpower Studies, Maxwell Al, June 2001, p. 68.
32 David Fulghum, 'A Crowded Room', 2006, pp. 52–54.
33 Simpson and Simpson, 2006, p. 10.
34 Cortes, 2003, p. 1.
35 Lok, 1 March 2004.
36 Ernie Regehr, *Canada and Ballistic Missile Defence*, Vancouver: Liu Institute for Global Issues, December 2003, p. 37.
37 Department of National Defence, *Backgrounder: Enhanced Canada–U.S. Defence Cooperation and the Bi-national Planning Group*, BG-04.041 – 1 April 2006. Available at www.forces.gc.ca/site/newsroom/view_news_e.asp?id=1528.
38 Bi-national Planning Group, *The Final Report on Canada and the United States (CANUS) Enhanced Military Cooperation*, 13 March 2006, p. 8.
39 Dwight Mason, 'Managing North American Defence at Home', paper presented at *What Canadian Military and Security Forces in the Future World? A Maritime Perspective*, Dalhousie University, 10–12 June 2005.
40 Colin S. Gray, 'Canada and NORAD: A Study in Strategy', *Behind the Headlines*, vol. XXXI, nos 3–4, June 1972, p. 3.
41 Interview, MGen. Angus Watt (CF), National Defence Headquarters, Ottawa, 28 September 2005. Emphases made by the interviewee in the conversation.
42 Interview, Capt. (N) Kendall Card (USN) and Capt. (N) Richard Bergeron, (CF), NORAD Headquarters, 3 October 2005.
43 Interview, LGen. Eric Findley, (CF), NORAD Headquarters, 3 October 2005.
44 Presentation by LGen. Charles Bouchard, DCOMNORAD, to National Security Studies Programme, Canadian Embassy, Washington DC, 4 April 2008. Comments used with permission.
45 Philippe Lagassé, 'Northern Command and the Evolution of Canada–US Defence Relations', *Canadian Military Journal*, Spring 2003, p. 16.
46 Jack English, *National Policy and the Americanization of the Canadian Military*, DCIEM Report # CR 2001–048, April 2001, p. 33.
47 Joseph Jockel, *No Boundaries Upstairs*, Vancouver: University of British Columbia Press, 1987, p. 17.
48 Anne Denholm-Crosby, *Dilemmas in Defence Decision Making: Constructing Canada's Role in NORAD 1958–1996*, New York: St. Martin's Press, 1998, pp. 30–31.
49 Joseph Jockel, *Canada in NORAD, 1957–2007: A History*, Kingston: Queen's University Centre for International Relations and the Queen's Defence Management Program, 2007, pp. 22–25.
50 Joseph Jockel, *No Boundaries Upstairs*, 1987, p. 4.
51 '5 Years After 9/11 – A CANR Perspective', *CCN Mathews Newswire*, 9 September 2006, p. 1.
52 Bi-national Planning Group, 2006, p. 2.

53 Dwight Mason, 'Time to Expand NORAD', *Security and Sovereignty: Renewing NORAD, One Issue, Two Voices #3*, Woodrow Wilson International Center for Scholars, 2005, p. 3.
54 Bi-national Planning Group, 2006, p. C1.
55 Jockel, *Canada in NORAD, 1957–2007*, 2007, p. 29.
56 Interview LGen. Findley (CF), NORAD Headquarters, 3 October 2005; confidential interview, NORAD Headquarters, 3 October 2005.
57 Bi-national Planning Group, 2006, p. C-2.
58 DiPasquale, 'NORAD, StratCom Linked on Air and Space Architecture', 2004; 'Combatant Commanders' Integrated Command and Control System (CCIC2S)', *Jane's C4I Systems*, 2007.
59 Interview, Capt. (N) Kendall Card (USN), Capt. (N) Richard Bergeron (CF), NORAD Headquarters, 3 October 2005.
60 Amy Butler and David Fulghum, 'F-15s Grounded Around the World', *Aviation Week and Space Technology*, 12 November 2007.
61 Bruce Rolfsen, '191 F-15s Grounded at Least Another Month', *Air Force Times*, 11 January 2008.
62 Bruce Campion-Smith, 'NORAD Facing "Rogue Elements" US General Says', *Toronto Star*, 10 April 2008; Mjr. Paul Doyle and Capt. William Mitchell, '425 Squadron Patrols the Alaska NORAD Region', *3 Wing News and Events*, available at www.airforce.forces.gc.ca/3wing/news/releases_e.asp?cat=26&id=5698.
63 Interview with MGen. Angus Watt (CF), National Defence Headquarters, Ottawa, 28 September 2005.
64 Presentation by LGen. Charles Bouchard, DCOMNORAD, to the National Security Studies Programme, Canadian Embassy, Washington DC, 4 April 2008. Comments used with permission.
65 Anne Denholm-Crosby notes, for example, 'On the basis of prior and privileged access to continental air defence planning, the Canadian military then controlled and manipulated the flow of information to the Canadian government as it deliberated the issues.' Denholm-Crosby, 1998, p. 34.
66 Denholm-Crosby, 1998, p. 28.
67 James M. Minifie, *Peacemaker or Powdermonkey: Canada's Role in a Revolutionary World*, Toronto: McClelland & Stewart, 1960, p. 99.
68 Michael Byers, 'Canadian Armed Forces Under United States Command', *International Journal*, vol. 58, no. 1, Winter 2002–2003, p. 89.
69 As Phillipe Lagassé notes, despite the outrage of Diefenbaker over not being consulted to the level he thought appropriate, the Cuban Missile Crisis demonstrates that the US did *not* exercise command over the Canadian Forces, this at the most dangerous moment in Cold War history. The functional ability and the political rationale for such control has declined significantly since 1962, rendering any future threat of this nature effectively moot, no matter how attractive a tar baby it remains for Canadian nationalists. Phillipe Lagassé, 'Tradition and Isolation: Canada, NorthCom and the UCP', p. 9, available at www.cda-cdai.ca/symposia/2002/lagasse.htm.
70 Jockel, *Canada in NORAD, 1957–2007*, 2007, pp. 30–31.
71 Ibid., p. 188.
72 Gray, 'Canada and NORAD: A Study in Strategy', 1972, p. 9. Emphasis added.
73 Jockel, *Canada in NORAD, 1957–2007*, 2007, p. 36.
74 Gray, 'Canada and NORAD: A Study in Strategy', 1972, p. 4.
75 One Canadian defence analyst argues that the different strategic approaches of each nation are evident in the histories discussing NORAD; American histories focus on the technical developments affecting NORAD, while Canadian histories concern themselves with the continuous tugs on that nation's sovereign independence. Martin Shadwick, 'NORAD, Sovereignty, and Changing Technology', *YCISS Occasional Paper #3*, Toronto: York Centre for International and Strategic Studies, 1985, pp. 1–2.

76 Jockel, *Canada in NORAD, 1957–2007*, 2007, pp. 131–132.
77 Joseph Jockel, 'Four US Military Commands: NORTHCOM, NORAD, SPACECOM, STRATCOM', *Institute for Research in Public Policy Working Paper # 2003–03*, p. 5.
78 Kevin Johnson, *The Effect of Command Structures on Canada's Participation in NORAD and ACLANT*, Masters thesis, Department of Political Science, University of Calgary, 1991, pp. 111–112.
79 Jockel, *Canada in NORAD, 1957–2007*, 2007, pp. 131, 136–137. Jockel notes that these restrictions were later lifted; e-mail to author, 15 July 2008.
80 Jockel, 'Four US Military Commands', pp. 3–5.
81 Denholm Crosby, 1998, pp. 1–2.
82 James Fergusson, 'NORAD Renewal – Much Ado About…', *Security and Sovereignty: Renewing NORAD, One Issue, Two Voices #3*, Woodrow Wilson International Center for Scholars, 2005, p. 9.
83 Lagassé, 'Tradition and Isolation: Canada, NorthCom and the UCP', p. 13.
84 Joseph T. Jockel and Joel J. Sokolsky, *The End of the Canada US Defence Relationship*, Kingston: Centre for International Relations Queen's University, May 1996.
85 Jeremy Feiler, 'US/Canada Could Swap Missile Defence Diplomatic Notes This Week', *Inside the Pentagon*, 15 January 2004.
86 Jeremy Feiler, 'US–Canada to Begin Negotiations on Common BMD', *Inside Missile Defense*, 21 January 2004; DiPasquale, 'Canada to Keep Traditional Missile Defense Role Under New US Plan', 2004.
87 David Szabo and Todd M. Walters (eds) *The Canada US Partnership: Enhancing our Common Security. Workshop Report*, Washington DC: Institute for Foreign Policy Analysis, 2005, p. 9, available at www.ifpa.org/pdf/Canada-US-Report.pdf.
88 Confidential interview, NORAD Headquarters, 3 October 2005; interview MGen. Angus Watt (CF), National Defence Headquarters, Ottawa 28 September 2005.
89 Fergusson, 'NORAD Renewal – Much Ado About…', 2005, p. 10.
90 Jockel and Sokolsky, 'Renewing NORAD Now If Not Forever', 2006, pp. 54–55.
91 Lagassé, 'Tradition and Isolation: Canada, NorthCom and the UCP', p. 21.
92 Jockel, *Canada in NORAD, 1957–2007*, 2007, p. 146.
93 Jockel, *No Boundaries Upstairs*, 1987, pp. 96, 107.
94 'In Brief – NATO Ends North America Deployment', *Jane's Defence Weekly*, 22 May 2002.
95 Otto Kreisler, 'The Years of Noble Eagle', *Air Force Magazine*, vol. 90, no. 6, June 2007.
96 Jockel, *Canada in NORAD, 1957–2007*, 2007, pp. 169–170.
97 William B. Scott, 'Exercise Jump-Starts Response to Attacks', *Aviation Week and Space Technology*, 3 June 2002. Available at www.aviationnow.com/content/publication/ awst/20020603/avi_stor.htm.
98 Retired House Representative Lee Hamilton (D-IN) noted that 'failure to share information is one cause of the severity of the September 11 attacks'. Libby John, '9/11 Commission issues below average grades for Information Sharing', *Inside the Air Force*, 9 December 2005.
99 DiPasquale '9/11 Commission Finds NORAD in '01 Weak on C^2 and Communications', 2004.
100 John T. Bennett, 'Notes From the Information Superiority Conference, July 19–20; Washington DC', *Inside the Air Force*, 22 July 2005.
101 Bruce Campion Smith, 'NORAD Facing "Rogue Elements" US General Says', *Toronto Star*, 10 April 2008.
102 Mason, 'Managing North American Defence at Home', June 2005, pp. 1–3.
103 Bernard Stancati, 'The Future of Canada's Role in Hemispheric Defense', *Parameters*, Autumn 2006, p. 104.

104 Fergusson, 'NORAD Renewal – Much Ado About…', 2005, p. 11.
105 Paul T. Mitchell, '1000 Ship Navies, Maritime Domain Awareness, and Networks: The Policy Nexus', *RUSI Defence* Systems, vol. 10, no. 1, June 2007, p. 65; Aarti Shah, 'McHale: Maritime NORAD Should be More Than Equivalence of Air Model', *Inside the Navy*, 13 June 2005; Zachery Petersen 'NORAD Beginning to Develop Plan for New Maritime Warning Mission', *Inside the Pentagon*, 5 July 2006; Emelie Rutherford, 'Officials Aim to Improve Global Maritime Situational Awareness', *Inside the Navy*, 16 April 2007.
106 Mike Blanchfield, 'Canada Kept in the Loop at NORAD About All Missile Threats', *Ottawa Citizen*, 10 April 2008.
107 Sebastien Sprenger, 'US, Canadian Troops Could Respond Jointly to Terrorist Attacks', *Inside the Army*, 5 September 2005.
108 Interview, Capt. (N) Kendall Card (USN) and Capt. (N) Richard Bergeron (CF), NORAD Headquarters, 3 October 2005.
109 Lagassé, 'Tradition and Isolation: Canada, NorthCom and the UCP', p. 18.
110 Stancati, 2006, pp. 109–110.
111 Ibid., pp. 111–112.
112 Jockel and Sokolsky, 'Renewing NORAD – Now if not Forever', 2006, pp. 54–56; Jason Sherman, 'Bush Approves Updates to UCP, Assigns New Missions', *Inside the Army*, 5 June 2006.
113 Mason, 'Time to Expand NORAD', 2005, p. 2.
114 Mason, 'Managing North American Defence at Home', 2005, p. 12.
115 Bi-national Planning Group, 2006, p. 17.
116 DiPasquale, 'Command, NORAD CIO Focused on Future Technological Relevancy', 2004.
117 Bi-national Planning Group, 2006, pp. C-2–C-5.
118 Interview, LGen. Eric Findley, (CF), NORAD Headquarters, 3 October 2005.
119 Confidential interview, NORAD Headquarters, 3 October 2005.
120 Mason, 'Time to Expand NORAD', 2005, p. 5.
121 Jockel and Sokolsky, 'Renewing NORAD – Now if not Forever', 2006, p. 57.
122 Bi-national Planning Group, 2006, p. 23.
123 Mason, 'Time to Expand NORAD', 2005, p. 4.
124 Bi-national Planning Group, 2006, p. C-6.
125 Jockel and Sokolsky, 'Renewing NORAD – Now if not Forever', 2006, pp. 53–54.
126 Stancati, 2006, pp. 103–104.
127 Mason, 'Time to Expand NORAD', 2005, p. 6.
128 Fergusson, 'NORAD Renewal – Much Ado About…', 2005, p. 12.
129 Interview LGen. Eric Findley (CF), NORAD Headquarters, 3 October 2005.
130 Interview Capt. (N) Kendall Card (USN) and Capt. (N) Richard Bergeron, (CF), NORAD Headquarters, 3 October 2005.
131 Interview with MGen. Angus Watt (CF), National Defence Headquarters, Ottawa, 28 September 2005.
132 Confidential interview, NORAD Headquarters, 3 October 2005.
133 Interview with Col. Eric Stevens (CF), NORAD Headquarters, 3 October 2005.
134 Glen Butler, 'Noble Eagle is not Your Average Operation', *Proceedings*, vol. 129, no. 8, August 2003, p. 48.
135 Jockel and Sokolsky, 'Renewing NORAD – Now if not Forever', 2006, pp. 57–58.
136 Szabo and Walters, 2005, p. 13.
137 Marina Malenic, 'Allies May Want Role in Missile Defense Command and Control', *Inside the Army*, 21 May 2007.
138 'NORAD East?', *Inside the Pentagon*, 17 May 2007.
139 Malenic, 2007.

6 Information, geography, mobility, and coordination: land operations in digital coalition battlespaces

1 Giles Ebbut, 'UK Command and Control during Iraqi Freedom', *Jane's Defence Weekly*, 1 July 2003.
2 US Department of Defense, *Transformation Planning Guidance*, Washington DC: US Department of Defense, April 2003, p. 3.
3 Ibid., pp. 11–12.
4 President George W. Bush, remarks to The Citadel, Charleston, SC, 11 December 2001.
5 Douglas A. MacGregor, *Breaking the Phalanx*, Westport, CN: Praeger, 1997, p. 44.
6 Lt. Col. James Boling, 'Rapid Decisive Operations: The Emperor's New Clothes of Modern Warfare', in Williamson Murray (ed.), *Transformation Concepts*, Carlisle PA: Strategic Studies Institute, 2002, pp. 167–168.
7 David Jablonsky, 'A Tale of Two Doctrines', in Conrad C. Crane, (ed.), *Transforming Defense*, Carlisle PA: Strategic Studies Institute, 2001, p. 46.
8 Jablonsky notes that the problems of Task Force Hawk were at the time well understood by the Army. He points to the 'High Technology Light Division' initiative of the 1980s, 1992's 'Louisiana Manoeuvres', and the 'Force XXI' initiative all as examples of the Army's attempt to increase the deployability of its forces without sacrificing firepower or protection. Ibid., pp. 47–48.
9 MacGregor, 1997, p. 37.
10 Lt. Col. H.R. McMaster (USA), 'Crack in the Foundation: Defense Transformation and the Underlying Assumption of Dominant Knowledge in Future War', *CSL Student Issue Paper*, Vol. S03–03, November 2003, p. 56.
11 Ibid., p. 57.
12 MacGregor, 1997, pp. 50–52.
13 Ibid., p. 54.
14 Thomas K. Adams, *The Army After Next: The First Post-Industrial Army*, Westport CN: Praeger Security International, 2006, p. 184.
15 Congressional Budget Office, *The Army's Future Combat Systems Program and Alternatives*, Washington DC: US Congressional Budget Office, 2006, pp. 21–26.
16 'New Landwarrior System Digitizes the Battlefield', *Spacewar News*, 15 January 2007.
17 Lisa Troshinsky, 'Force Networking in Operation Iraqi Freedom a Big Improvement, Official Says', *Aerospace Daily*, vol. 209, no. 1, 26 January 2004, p. 3.
18 See, Mark C. Malham and Deborah Gabbard, 'Battle Command Systems: The Force XXI Warfighter's Advantage', *Military Review*, vol. 78, no. 2, March/April 1998; 'Army Tactical Command and Control System', www.fas.org/man/dod-101/sys/land/atccs.htm.
19 Michele Zanini and Jennifer Morrison Taw, *The Army and Multinational Force Compatibility*, Santa Monica: Rand Arroyo Center, 2000, pp. 14–15.
20 Wing Commander H. Smythe (RAF), 'From Coningham to Project Coningham Keyes', student paper, Joint Command and Staff College, Watchfield UK, p. 22.
21 Anthony Cordesman, *The Iraq War: Strategy Tactics and Military Lessons*, Westport CN: Praeger Security International, 2004, p. 216.
22 Richard J. Dunn III, *Blue Force Tracking: The Afghanistan and Iraq Experience and Its Implications for the US Army*, Reston Virginia: Northrop Grumman Mission Systems, 2003, p. 12.
23 Dennis Murphy, 'Networked Enabled Operations: Initial Impressions', *CSL Issue Papers*, Vol. 06–05, March 2005, p. 3.
24 Cordesman, 2004, p. 349.
25 Imad Bitar and Brian L. Felsman. 'Blue Force Tracker in OEF and OIF', *Technology Journal*, Fall/Winter 2005, p. 81; Dunn, 2003, p. 6; 'Force XXI Battle Command Brigade and Below', www.ms.northropgrumman.com/markets/MDFbcb2.html.

26 Dunn, 2003, p. 3.
27 Malham and Gabbard, March/April 1998.
28 See especially John W. Charlton. 'Digital Command: Baptism by Fire', *Armor*, November 2003.
29 Malham and Gabbard, March/April 1998.
30 Michael Knights, *Cradle of Conflict: Iraq and the Birth of Modern US Military Power*, Annapolis: Naval Institute Press, 2005, pp. 100–101, 114.
31 Dunn, 2003, pp. 9–12.
32 'Afghanistan and Iraq Test Theory of Network Centric Warfare', *Federal Computer Week*, vol. 21, no. 2, 22 January 2007, p. S2.
33 Ann Roosevelt, 'Army Considers Improving Oneness of FBCB2 Systems', *C4I News*, 5 February 2004, p. 1.
34 Hunter Keeter, '4th Infantry Division Leverages Technology across its Mission Spectrum', *C4I News*, 24 June 2003, p. 1.
35 Dunn, 2003, p. 6.
36 McMaster, pp. 33–37. See also Adams, 2006.
37 Mark Unewisse, Paul Gaertner, Anne Marie Grisogono and Robert Seymore, 'Land Situational Awareness for 2010', Defence Science and Technology Organisation paper, Australian Defence Force, p. 2. Available at www.siaa.asn.au/get/2395379372.pdf.
38 Curtis Taylor, 'Trading the Saber for Stealth – Can Surveillance Technology Replace Traditional Aggressive Reconnaisance?', *The Land Warfare Papers*, no. 53, September 2005, pp. 14–16.
39 Josh Kucera, 'Red Force Tracking Advances', *Jane's Defence Weekly*, 4 March 2005; Josh Kucera, 'US Surveillance Link-up Will Boost Ability to Track the Enemy', *Jane's Defence Weekly*, 24 March 2005.
40 Cordesman, 2004, p. 366.
41 Ibid., p. 227.
42 McMaster, p. 19.
43 Taylor, 2005, p. 8.
44 Ibid., pp. 3–7.
45 Maj. J. Anderson (BA), 'A Network Enabled Capability: How will it Change the Way we Operate?', Student paper, Advanced Command and Staff Course, Joint Services Command and Staff College, p. 18.
46 McMaster, 2003, p. 9.
47 Brian Robinson, 'DOD Arms Soldiers, Allies with Information', *Federal Computer Week*, vol. 20, no. 25, 31 July 2006.
48 Lt. Col. John D. Nelson (USA), 'Swiftly Defeat the Efforts: Then What?', *CSL Student Issue Paper*, Vol. S04–04, July 2004, pp. 1–2, 5–6.
49 W. H. Moore, 'The United Kingdom's View of US Army Transformation', in Conrad C. Crane, (ed.), *Transforming Defense*, Carlisle PA: Strategic Studies Institute, 2001, p. 169.
50 Karl Ritter, 'Report: Risk of Nuclear Warfare Rising', *Washington Post*, 11 June 2007, www.washingtonpost.com/wp-dyn/content/article/2007/06/11/AR2007061100663.html.
51 Ministry of Defence, *Network Enabled Capability*, London: HMSO, 2002, pp. 2–3.
52 Maj. John Owens (BA), 'NEC: How Will it Affect the Philosophy of Mission Command?', Student paper, Advanced Command and Staff Programme, Joint Services Command and Staff College, p. 3.
53 Owens argues that self-synchronisation is the 'ability of a well informed force to organise and synchronise complex warfare activities from the bottom up. The organising principles are unity of effort, clearly articulated commander's intent, and carefully crafted ROE.' Mission Command, on the other hand is 'a style of combat which provides decentralised command, freedom and speed of action, and initiative'. Commanders provide the intent of the mission and its context, subordinates are told which

tasks to achieve within allocated resources. A minimum of control is exerted over subordinates, who decide the best means for achieving their mission. The concepts may differ to a degree, but the distinction is a narrow one at best. Ibid., pp. 3, 6.

54 Anderson, p. 4.
55 Moore, 2001, pp. 170–172.
56 Christopher F. Foss, 'FRES Project Gathers Speed', *Jane's Defence Weekly*, 20 April 2005.
57 Peter Felstead, 'AA2006: FRES gets Heavier – Armoured Doctrinal Rethink', *Jane's Defence Weekly*, 1 March 2006.
58 Written Ministerial Statements for 5 May 2004. House of Lords Hansard, www.publications.parliament.uk/pa/ld200304/ldhansrd/vo040505/text/40505 12.htm.
59 National Audit Office, *Delivering Digital Tactical Communications through the Bowman CIP*, London: HMSO, 2006, pp. 7–8.
60 Tim Ripley, 'Part One: Communications – Speak Easy', *Jane's Defence Weekly*, 27 April 2005.
61 National Audit Office, *Delivering Digital Tactical Communications through the Bowman CIP*, London: HMSO, 2006, p. 2.
62 Ebbut, 'UK Command and Control during Iraqi Freedom', 2003.
63 Office of Force Transformation, *US/UK Coalition Combat Operations during Operation Iraqi Freedom*, Washington DC: Department of Defense, 2005, pp. 6–3; Ebbut, 'UK Command and Control during Iraqi Freedom', 2003. Ebbut remarks that the UK was also able to access the SIPRNET through a network 'tunnel' known as X-NET.
64 Ibid.
65 Ibid.
66 Jim Dutton and Tom Waldhouser, 'US–UK Operations', *RUSI Journal*, December 2003, p. 12; UK-US Operations, pp. 5–2, 3–5.
67 See: Ripley, 2005.
68 Dutton and Waldhouser, 2003, p. 12.
69 Tim Ripley, 'Major Flaws in UK's Iraq War Communications', *Jane's Defence Weekly*, 7 January 2004, p. 13.
70 Office of Force Transformation, 2005, pp. 6–2.
71 Jefferson Morris, 'UK Likely to Provide More Funding for DOD Sponsored NCW Study', *Aerospace Daily*, vol. 209, iss. 12, 22 June 2004, p. 1.
72 Smythe, pp. 18–19.
73 Indeed, the same technological demands may actually *prevent* a coalition partner from supplying close air support at all, even perhaps to its own forces. One study notes that UK forces are 'incapable of achieving combat identification at medium altitudes', suggesting that the RAF, for fear of fratricide, may be as unwelcome in a digitised environment as 'analogue' ground formations.
74 Maj. Jack L. Sine II (USA), *Organising the Fight: Technological Determinants of Coalition Command and Control and Combat Operations*, MA thesis, Naval Postgraduate School, September 2006, pp. 13, 17.
75 Darren Lake and Kim Burger, 'UK, USA Must do More to Reduce Friendly Fire Risk', *Jane's Defence Weekly*, 22 January 2003.
76 Adams, 2006, p. 209.
77 Maj. Jack L. Sine II (USA), *Organising the Fight: Technological Determinants of Coalition Command and Control and Combat Operations*, pp. 20–22.
78 Office of Force Transformation, *US/UK Coalition Combat Operations during Operation Iraqi Freedom*, Washington DC: Department of Defense, 2005, pp. 5–2.
79 Zanini and Taw, 2000, pp. 22–23.
80 Ibid., p. 30.
81 Sine, 2006, pp. 32–33.
82 Ibid., 2006, p. 61.

83 'Kreig Reaffirms Imperative for Coalition Interoperability, but Says Process will Take Time', *Defense Daily International*, vol. 7, no. 35, 8 September 2006, p. 1.
84 Paul T. Mitchell, 'International Anarchy and Coalition Interoperability in High Tech Environments', in David Carment, Martin Rudner (eds), *Peacekeeping Intelligence: New Players, Extended Boundaries*, Abingdon: Routledge, 2005.
85 Dunn, 2003, p. 15.

Conclusion

1 Alan Ryan, 'Australian Army Cooperation with the Land Forces of the United States: Problems of a Junior Partner', *Land Warfare Studies Centre Working Paper*, no. 121, January 2003, p. 34.

Bibliography

Anonymous, 'Force XXI Battle Command Brigade and Below', www.ms.northropgrumman.com/markets/MDFbcb2.html.

—— 'In Rumsfeld's Words: Guidelines for Committing Forces', *New York Times*, 14 October 2002.

—— 'In Brief – NATO Ends North America Deployment', *Jane's Defence Weekly*, 22 May 2002.

—— 'Wikipedia Study "Fatally Flawed"', *BBC News*, http://news.bbc.co.uk/2/hi/technology/4840340.stm.

—— 'General Warns over Digitization Split', *International Defence Review*, 1 January 2002.

—— *Griffin Key Attributes*, 25 January 2005, www.jcs.mil/j6/cceb/griffinkeyattributes26jan05.pdf.

—— 'Internet Encyclopaedias Go Head to Head', *Nature*, 15 December 2005, www.nature.com/nature/journal/v438/n7070/full/438900a.html.

—— 'Kreig Reaffirms Imperative for Coalition Interoperability, but says process will take time', *Defense Daily International*, vol. 7, no. 35, 8 September 2006.

—— '5 Years After 9/11 – A CANR Perspective', *CCN Mathews Newswire*, 9 September 2006.

—— 'Combatant Commanders' Integrated Command and Control System (CCIC2S)', *Jane's C4I Systems*, 2007.

—— 'New Landwarrior System Digitizes the Battlefield', *Spacewar News*, 15 January 2007.

—— 'Afghanistan and Iraq Test Theory of Network Centric Warfare', *Federal Computer Week*, vol. 21, no. 2, 22 January 2007.

—— 'NORAD East?', *Inside the Pentagon*, 17 May 2007.

Barbara Adam, Ulrich Beck, and Joost van Loon, *Risk Society and Beyond: Critical Issues for Social Theory* (London: Sage, 2000).

Charlotte Adams, 'Network Centric Rush To Connect', *Aviation Today*, 1 September 2004.

Cpt. Marcella Adams (USAF), 'Controlling the Bosnian Skies', *Airman*, available at www.af.mil/news/airman/0896/caoc.htm.

Thomas K. Adams, *The Army After Next: The First Post-Industrial Army* (Westport CN: Praeger Security International, 2006).

Davi M. D'Agostino, *Managing Sensitive Information: DoD Can More Effectively Reduce the Risk of Classification Errors* (Washington DC: Government Accounting Office, June 2006).

Robert Akerman, 'Aerospace Experts Refocus the Tactical Picture', *Signal*, vol. 55, no. 3, November 2000.

David S. Alberts, John J. Gartska, and Frederick P. Stein, *Network Centric Warfare: Developing and Leveraging Information Superiority*, 2nd edition (Washington DC: Command and Control Research Program, 1999).

David S. Alberts, John J. Gartska, Richard E. Hayes, and David A. Signori, *Understanding Information Age Warfare* (Washington DC: Command and Control Research Program, 2001).

David S. Alberts and Richard E. Hayes, *Power to the Edge: Command and Control in the Information Age* (Washington DC: Command and Control Research Program, 2003).

Chris Anderson, 'Free! Why $0.00 Is the Future of Business', *Wired*, 25 February 2008, www.wired.com/techbiz/it/magazine/16–03/ff_free.

Maj. J. Anderson, 'A Network Enabled Capability: How will it Change the Way we Operate?', Student paper, Advanced Command and Staff Course, Joint Services Command and Staff College.

Duane P. Andrews (chairman), *Report of the Defense Science Board Task Force on Information Warfare Defense* (Washington DC: Defense Science Board, November 1996).

Calvin Andrus, 'The Wiki and the Blog: Toward an Adaptive Intelligence Community', *Studies in Intelligence*, vol. 49, no. 3, September 2005, available at http://ssrn.com/abstract=755904.

Edwin Leigh Armistead, *AWACS and Hawkeyes* (St Paul, MN: MBI Publishing, 2002).

Raymond Aron, *Peace and War: A Theory of International Relations* (New York: Doubleday, 1966).

Maria Aspan, 'Ease of Alteration Creates Woes for Picture Editors', *New York Times*, 14 August 2006, www.nytimes.com/2006/08/14/technology/14photoshop.html.

Jonathan B. A. Bailey, 'The First World War and the Birth of Modern Warfare', in MacGregor Knox and Williamson Murray (eds), *The Dynamics of Military Revolution* (Cambridge: Cambridge University Press, 2001).

Paul Baran, *On Distributed Communications: IX. Security, Secrecy, and Tamper Free Considerations* (Santa Monica: Rand Corporation, 1964).

John Perry Barlow, 'The Economy of Ideas', *Wired*, 2 March 1994, www.wired.com/wired/archive/2.03/economy.ideas.html.

Thomas P. Barnett, *The Pentagon's New Map: War and Peace in the Twenty-First Century* (New York: Berkley Books, 2004).

Ulrich Beck, *Risk Society: Towards a New Modernity* (London: Sage, 1992).

—— *World Risk Society* (Cambridge: Polity Press, 1999).

John T. Bennett, 'Notes From the Information Superiority Conference, July 19–20; Washington DC', *Inside the Air Force*, 22 July 2005..

Richard Best, 'Sharing Law Enforcement and Intelligence Information: The Congressional Role', Congressional Research Service Paper RL33873, 13 February 2007.

Nikolai Bezroukov, 'A Second Look at the Cathedral and the Bazaar', *First Monday*, vol. 4, no. 12, December 1999, www.uic.edu/htbin/cgiwrap/bin/ojs/index.php/fm/article/view/708/618.

Bi-national Planning Group, *The Final Report on Canada and the United States (CANUS) Enhanced Military Cooperation*, 13 March 2006.

Imad Bitar and Brian L. Felsman, 'Blue Force Tracker in OEF and OIF', *Technology Journal*, Fall/Winter 2005.

Major Michael B. Black (USA), 'Coalition Command, Control, Communications, Computer and Intelligence Systems Interoperability: A Necessity or Wishful Thinking?', thesis, US Army Command and General Staff College, Fort Leavenworth, KS, 2 June 2000.

Mike Blanchfield, 'Canada Kept in the Loop at NORAD About All Missile Threats', *Ottawa Citizen*, 10 April 2008.

Max Blumenthal, 'Data Debase', *American Prospect*, 19 December 2003, www.prospect.org/webfeatures/2003/12/blumenthal-m-12–19.html.

Lt. Col. James Boling (USA), 'Rapid Decisive Operations: The Emperor's New Clothes of Modern Warfare', in Williamson Murray (ed.), *Transformation Concepts* (Carlisle PA: Strategic Studies Institute, 2002).

British Parliament, Written Ministerial Statements for 5 May 2004. House of Lords Hansard, www.publications.parliament.uk/pa/ld200304/ldhansrd/vo040505/text/40505–12.htm.

Alice R. Buchalter, John Gibbs and Marieke Lewis, *Laws and Regulations Governing the Protection of Sensitive but Unclassified Information* (Washington DC: Library of Congress, September 2004).

Hedley Bull, *The Anarchical Society, A Study of Order in World Politics* (London: MacMillan, 1977).

Robert Burnett and P. David Marshall, *Web Theory: An Introduction* (London: Routledge, 2003).

George W. Bush, remarks to The Citadel, Charleston, S.C., 11 December 2001.

—— *The National Security Strategy of the United States*, 2002, www.whitehouse.gov/nsc/nss5.html.

Amy Butler and David Fulghum, 'F-15s Grounded Around the World', *Aviation Week and Space Technology*, 12 November 2007.

Commander John Bycroft (CF), 'Coalition C4ISTAR Capability AUSCANUKUS', unpublished paper presented to the SMi conference 'Naval C4ISTAR', London, 21 April 2004.

Michael Byers, 'Canadian Armed Forces Under United States Command', *International Journal*, vol. 58, no. 1, Winter 2002–2003.

Frances Cairncross, *The Death of Distance* (Boston: Harvard Business School Press, 1997).

David Calleo, 'Power, Wealth, and Wisdom: The United States and Europe after Iraq', *The National Interest*, Summer 2003.

Steven Cambone, 'Memorandum: Security Classification Marking Instructions', 27 September 2004.

Bruce Campion-Smith, 'NORAD Facing "Rogue Elements" US General Says', *Toronto Star*, 10 April 2008.

Colonel Thomas A. Cardwell (USAF), *Airland Combat: An Organization for Joint Warfare* (Maxwell, AL: Air University Press, 1992).

E. H. Carr, *The Twenty Years Crisis 1919–1939* (New York: Harper and Row, 1964).

Commander James Carr, 'Network Centric Coalitions: Pull, Pass, or Plug-in?', course paper, Naval War College, Newport, RI, 15 May 1999.

Manuel Castells, *End of the Millenium* (Malden MA: Blackwell, 1998).

—— *The Rise of the Network Society, Second Ed.* (Malden MA: Blackwell, 2000).

—— 'The Network Society', in Pekka Himanen (ed.), *The Hacker Ethic* (New York: Random House, 2001).

—— 'Informationalism, Networks and the Network Society: A Theoretical Blueprint', in Manuel Castells (ed.), *The Network Society: A Cross-Cultural Perspective* (Cheltenham: Edgar Elgar, 2004).

John W. Charlton, 'Digital Command: Baptism by Fire', *Armor*, November 2003.

Colonel Robert Chekan, 'The Future Of Warfare: Clueless Coalitions?', course paper, Canadian Forces College, October 2001.

Ian Clark, *Globalization and Fragmentation: International Relations in the Twentieth Century* (Oxford: Oxford University Press, 1997).

Carl von Clausewitz, *On War* (Princeton, NJ: Princeton University Press, 1976).

Kathryn Cochrane, 'Kosovo Targeting – A Bureaucratic and Legal Nightmare', *Aerospace Centre Paper 3* (Canberra: Aerospace Development Centre, June 2001).

Eliot Cohen, 'Change and Transformation in Military Affairs', *Journal of Strategic Studies*, vol. 27, no. 3, September 2004.

Christopher Coker, *Waging Wars Without Warriors: The Changing Culture of Military Conflict* (London: IISS, 2002).

—— *Globalisation and Insecurity in the Twenty-first Century: NATO and the Management of Risk, Adelphi Paper #345* (London: IISS, 2002).

—— *The Future of War: The Re-enchantment of War in the Twenty-first Century* (London: Blackwell Manifestos, 2004).

Major Robert L. Coloumbe (USMC), 'Operational Art and NATO C4I: An Oxymoron?', course paper, Department of Joint Military Operations, US Naval War College, Newport, RI, 5 February 2001.

Combined Communications Electronics Board, *A Strategy for Improved Coalition Networking*, June 2005, p. 1, www.jcs.mil/j6/cceb/cnsdatedjune05.pdf.

Committee on Network-Centric Naval Forces, Naval Studies Board, *Network Centric Naval Forces: A Transition Strategy for Enhancing Operational Capabilities* (Washington DC: National Academy Press, 2000).

Committee to Review DOD C4I Plans and Programs, Computer Science and Telecommunications Board, National Research Council, *Realizing the Potential of C4I* (Washington DC: National Research Council, 1999).

Congressional Budget Office, *The Army's Future Combat Systems Program and Alternatives* (Washington DC: US Congressional Budget Office, 2006).

Anthony Cordesman, *The Iraq War: Strategy Tactics and Military Lessons* (Westport CN: Praeger Security International, 2004).

Lorenzo Cortes, 'CAOC Crews Credit TBMCS and IWS for OIF Success', *C4I News*, 12 December 2003.

Robert F. Dacey, *Progress and Challenges to an Effective Defense-wide Information Assurance Program*, GAO-01–307 (Washington DC: GAO, March 2001).

Ivo Daalder and Michael O'Hanlon, *Winning Ugly* (Washington DC: Brookings, 2001).

Ron Deibert, John Palfrey, Rafal Rohinsky and Jonathon Zittrain, *Access Denied: The Practice and Policy of Global Internet Filtering* (Boston: Harvard University Press, 2008).

Anne Denholm-Crosby, *Dilemmas in Defence Decision Making: Constructing Canada's Role in NORAD 1958–1996* (New York: St. Martin's Press, 1998).

Department of the Army, *US Army Field Manual 100–5 Blueprint for the AirLand Battle* (Washington DC: Brassey's (US) Inc., 1991).

—— Army Regulation 530–1 'Operations and Signal Security' 19 April 2007, paragraph 2–1 g; available at http://blog.wired.com/defense/files/army_reg_530_1_updated.pdf.

Department of Defense, *Transformation Planning Guidance* (Washington DC: US Department of Defense, April 2003).

—— *Joint Publication 1–02, 'DOD Dictionary of Military and Associated Terms'*, www.dtic.mil/doctrine/jel/doddict/data/b/00700.html, as amended through 31 August 2005.

Department of National Defence, *Backgrounder: Enhanced Canada–U.S. Defence Cooperation and the Bi-national Planning Group*, BG-04.041 – 1 April 2006. Available at www.forces.gc.ca/site/newsroom/view_news_e.asp?id=1528.

Department of the Navy, SECNAV INSTRUCTION 5720.47B 'Department of the Navy Policy for Content on Publically Accessible World Wide Web Sites', 28 December 2005.

Cynthia DiPasquale, 'Command, NORAD CIO Focused on Future Technological Relevancy', *Inside the Air Force*, 16 January 2004.

—— 'NORAD, StratCom Linked on Air and Space Architecture', *Inside the Air Force*, 20 February 2004.

—— '9/11 Commission Finds NORAD in '01 Weak on C² and Communications', *Inside the Air Force*, 25 June 2004.

—— 'Canada to Keep Traditional Missile Defense Role Under New US Plan', *Inside the Air Force*, 27 August 2004.

Director of Central Intelligence Directive 1/7, 'Security Controls on the Dissemination of Intelligence Information', 15 June 1996, Sections 7 and 12, www.fas.org/irp/offdocs/dcid17m.htm.

Mjr. Paul Doyle and Capt. William Mitchell, '425 Squadron Patrols the Alaska NORAD Region', *3 Wing News and Events*, available at www.airforce.forces.gc.ca/3wing/news/releases_e.asp?cat=26&id=5698.

Richard J. Dunn III, *Blue Force Tracking: The Afghanistan and Iraq Experience and Its Implications for the US Army* (Reston Virginia: Northrop Grumman Mission Systems, 2003).

Jim Dutton and Tom Waldhouser, 'US UK Operations', *RUSI Journal*, December 2003.

Giles Ebbut, 'UK Command and Control during Iraqi Freedom', *Jane's Defence Weekly*, 1 July 2003.

Allan English, Richard Gimblett and Howard Coombs, *Beware of Putting the Cart Before the Horse: Network Enabled Operations as a Canadian Approach to Transformation*, DRDC Contract Report CR 2005–212, Toronto, 19 July 2005, p. 13, http://pubs.drdc-rddc.gc.ca/pubdocs/pcow1_e.html.

Jack English, *National Policy and the Americanization of the Canadian Military*, DCIEM Report # CR 2001–048, April 2001.

Jerry Everard, *Virtual States: The Internet and the Boundaries of the Nation-State* (London: Routledge, 2000).

Federation of American Scientists, 'Army Tactical Command and Control System', www.fas.org/man/dod-101/sys/land/atccs.htm.

Jeremy Feiler, 'US/Canada Could Swap Missile Defence Diplomatic Notes This Week', *Inside the Pentagon*, 15 January 2004.

—— 'US–Canada to Begin Negotiations on Common BMD', *Inside Missile Defense*, 21 January 2004.

Peter Felstead, 'AA2006: FRES gets Heavier – Armoured Doctrinal Rethink', *Jane's Defence Weekly*, 1 March 2006.

James Fergusson, 'NORAD Renewal – Much Ado About …', *Security and Sovereignty: Renewing NORAD, One Issue, Two Voices #3*, Woodrow Wilson International Center for Scholars, 2005.

Lieutenant-Colonel Chris Field (ADF), 'An Australian Defence Force Liaison Officer's Observations and Insights from Operation Iraqi Freedom', *Australian Defence Force Journal*, no. 163, November–December 2003.

Rosemary Foot, John Lewis Gaddis, and Andrew Hurrell (eds), *Order and Justice in International Relations* (Oxford: Oxford University Press, 2003).

Commander J. L. R. Foreman (RN), 'Multinational Information Sharing (MNIS)', unpublished briefing slides.

J. Franklin (ed.), *The Politics of Risk Society* (Cambridge: Polity, 1998).

Lawrence Freedman, 'Strategic Studies and the Problem of Power', in Lawrence Freedman, Paul Hayes and Robert O'Neill (eds), *War Strategy, and International Politics: Essays in Honour of Sir Michael Howard* (Oxford: Clarendon Press, 1992).

—— 'The Transatlantic Agenda: Vision and Counter-Vision', *Survival*, vol. 47, no. 4, Winter 2005–2006.

Norman Friedman, *The US Maritime Strategy* (London: Jane's Publishing, 1988).

—— *World Naval Weapons Systems 1997–1998* (Annapolis, MD: Naval Institute Press, 1997).

—— *The Fifty Year War: Conflict and Strategy in the Cold War* (Annapolis, MD: United States Naval Institute Press, 2000).

—— *US Destroyers Revised Edition* (Arlington, VA: Naval Institute Press, 2004).

Thomas Friedman, *The World is Flat* (New York: Farrar, Straus and Giroux, 2005).

Christopher F. Foss, 'FRES Project Gathers Speed', *Jane's Defence Weekly*, 20 April 2005.

David Fulghum, 'USAF Streamlines the Air Operations Center', *Aviation Week and Space Technology*, vol. 157, no. 13, 23 September 2002.

—— 'New Bag of Tricks', *Aviation Week and Space Technology*, vol. 158, no. 16, 2003.

—— 'A Crowded Room', *Aviation Week and Space Technology*, vol. 160, no. 17, 26 April 2006.

John Garnett, 'Limited War', in John Baylis, Ken Booth, John Garnett, and Phil Williams (eds), *Contemporary Strategy: Theories and Policies* (Beckenham: Croom Helm, 1975).

Kenneth Gause, 'US Navy Interoperability with Its High-End Allies', unpublished paper.

Barton Gellman, 'Pentagon Would Preclude a Rival Superpower', *Washington Post*, 11 March 1992.

Commander Barbara A. Geraghty (USN), 'Will Network Centric Warfare be the Death Knell for Allied/Coalition Operations?', course paper, Department of Joint Military Operations, US Naval War College, Newport, RI, 17 May 1999.

Eric Francis Germain, 'The Coming Revolution in NATO Maritime Command and Control', *Mitre Technical Papers*, www.mitre.org/support/papers/technet97/germain_technet.pdf.

Anthony Giddens, *The Consequences of Modernity* (Cambridge: Polity, 1990).

—— *Runaway World: How Globalisation is Reshaping our Lives* (London: Profile Books, 1999).

Richard Gimblett, *Operation Apollo* (Ottawa: Magic Light, 2004).

—— 'Command of Coalition Operations in a Multicultural Environment: A Canadian Naval Niche? The Case Study of Operation Apollo', unpublished paper prepared for the Canadian Forces Leadership Institute.

Globalsecurity.org, 'Automated Deep Operations Coordination System', www.globalsecurity.org/military/systems/ground/adocs.htm.

—— 'Global Information Grid (GIG)', www.globalsecurity.org/space/systems/gig.htm.

—— 'Global Information Grid (GIG) Bandwidth Expansion (GIG-BE)', www.globalsecurity.org/space/systems/gig-be.htm.

—— 'Theatre Battle Management Core Systems', www.globalsecurity.org/military/systems/aircraft/systems/tbmcs.htm.

—— 'Transformational Communications Architecture', www.globalsecurity.org/space/systems/tca.htm.

—— 'Transformational SATCOM (TSAT) Advanced Wideband System', www.globalsecurity.org/space/systems/tsat.htm.

James Goldric, 'In Command in the Gulf', *US Naval Institute Proceedings*, vol. 128, no. 12, December 2002.

Colonel George K. Gramer (USA), 'Optimizing Intelligence Sharing in a Coalition Environment: Why US Operational Commanders Have an Intelligence Dissemination Problem', course paper, Department of Joint Military Operations, US Naval War College, Newport, RI, 17 May 1999.

Colin S. Gray, 'Canada and NORAD: A Study in Strategy', *Behind the Headlines*, vol. XXXI, nos 3–4, June 1972.

—— *Strategy for Chaos* (London: Frank Cass, 2002).

Nicky Hager, *Secret Power: New Zealand's Role in the International Spy Network* (Nelson NZ: Craig Potton Publishing, 1996).

Victor David Hanson, *Carnage and Culture: Landmark Battles in the Rise of Western Power* (New York: Doubleday, 2001).

Mark Hewish, 'Out of CAOCs Comes Order', *International Defence Review*, 1 May 2003.

Pekka Himanen, 'The Hacker Ethic as the Culture of the Information Age', in Manuel Castells (ed.), *The Network Society: A Cross-Cultural Perspective* (Cheltenham: Edgar Elgar, 2004).

Frank G. Hoffman, 'The New Normalcy', *E-Notes*, www.fpri.org/enotes/20060512.americawar.hoffman.newnormalcy.html.

Michael Howard, 'Morality and Force in International Politics', *Studies in War and Peace* (London: Temple Smith, 1970).

Peter Howard, 'The USN's Designer of Concepts', *Jane's Defence Weekly*, 3 October 2001.

Justin Huggler, 'Israelis Trained US Troops in Jenin-Style Urban Warfare', *The Independent*, 29 March 2003.

Richard Hunter, *World Without Secrets: Business, Crime, and Privacy in the Age of Ubiquitous Computing* (New York: John Wiley and Sons, 2002).

Andrew Hurrell, 'Explaining the Resurgence of Regionalism in World Politics', *Review of International Studies*, vol. 21, no. 4, 1995.

Michael Ignatieff, *Virtual War: Kosovo and Beyond* (Toronto: Viking Press, 2000).

Lieutenant-Commander Ivan Ingham (RAN), 'Naval Gunfire Support for the Assault of the Al Faw Peninsular', *Journal of the Australian Naval Institute*, no. 109, Winter 2003.

David Jablonsky, 'A Tale of Two Doctrines', in Conrad C. Crane (ed.), *Transforming Defense* (Carlisle PA: Strategic Studies Institute, 2001).

LCdr. Jonathon Lee Jackson (USN), *Solving the Problem of Time Sensitive Targets*, Joint Military Operations Paper, US Naval War College, 3 February 2003.

Robert Jervis, *American Foreign Policy in a New Era* (New York: Routledge, 2005).

Joseph Jockel, *No Boundaries Upstairs* (Vancouver: University of British Columbia Press, 1987).

—— 'Four US Military Commands: NORTHCOM, NORAD, SPACECOM, STRATCOM', *Institute for Research in Public Policy Working Paper # 2003–03*.

—— *Canada in NORAD, 1957–2007: A History* (Kingston: Queen's University Centre for International Relations and the Queen's Defence Management Program, 2007).

Joseph T. Jockel and Joel J. Sokolsky, *The End of the Canada US Defence Relationship* (Kingston: Centre for International Relations Queen's University, May 1996).

—— 'Renewing NORAD Now if not Forever', *Policy Options*, July–August 2006.

Libby John, '9/11 Commission Issues Below Average Grades for Information Sharing', *Inside the Air Force*, 9 December 2005.

Kevin Johnson, *The Effect of Command Structures on Canada's Participation in NORAD and ACLANT*, Masters thesis, Department of Political Science, University of Calgary, 1991.

R. J. Barry Jones, *Globalisation and Interdependence in the International Political Economy: Rhetoric and Reality* (London: Pinter Publishers, 1995).

LCol. Joseph H. Justice III (USAF), *Air Power Command and Control: Evolution of the Air and Space Operations Center as a Weapon System*, US Army War College Research Project (Carlisle Barracks PA: US Army War College, 3 May 2004).

Frederick Kagan, 'The Military's Manpower Crisis', *Foreign Affairs*, vol. 85, no. 4, July–August 2006.

Robert Kagan and William Kristol, 'The Present Danger', *The National Interest*, Spring 2000.

Mary Kaldor, *New and Old Wars: Organised Violence in the Global Era* (Cambridge: Polity Press, 1999).

Sayaka Kawakami and Sarah C. McCartney, 'Government Information Collection: Privacy Year in Review: Privacy Impact Assessments, Airline Passenger Pre-Screening, and Government Data Mining', *I/S: A Journal of Law and Policy for the Information Society*, vol. 1, nos 2–3, Spring/Summer 2005.

David E. Kaplan and Kevin Whitelaw, 'Remaking US Intelligence', *US News and World Report*, 3 November 2006, www.usnews.com/usnews/news/articles/061103/3dni.intro.htm.

Johnathon Karp and Andy Pasztor, 'Pentagon Week: High Tech Has High Risk', *Wall Street Journal*, 2 May 2005, p. B2.

Hunter Keeter, '4th Infantry Division Leverages Technology across its Mission Spectrum', *C4I News*, 24 June 2003.

Johnny Kegler, 'Pathways to Enlightenment', *Armada International*, vol. 29, no. 5, October–November 2005.

Charles W. Kegley Jr. and Gregory A. Raymond, *When Trust Breaks Down: Alliance Norms and World Politics* (Columbia, SC: University of South Carolina, 1990).

Henry S. Kenyon, 'Alliance Forces Move Toward Unified Data Infrastructure', *Signal*, vol. 56, no. 1, September 2001.

James Kinniurgh and Dorothy Denning, 'Blogs and Military Information Strategy', *Joint Special Operations University Report 06–5*, June 2006.

Henry Kissinger, *The Troubled Partnership* (New York: McGraw Hill, 1965).

John Kiszely, 'Achieving High Tempo: New Challenges', *RUSI Journal*, vol. 144, no. 6, December 1999.

Michael Knights, *Cradle of Conflict: Iraq and the Birth of Modern US Military Power* (Annapolis: Naval Institute Press, 2005).

Mustafa R. Koprucu, *The Elements of Decentralized Execution: the Effect of Technology on a Central Air Power Tenet*, thesis prepared for the School of Advanced Airpower Studies, Maxwell Al, June 2001.

Otto Kreisler, 'The Years of Noble Eagle', *Air Force Magazine*, vol. 90, no. 6, June 2007.

Charles C. Krulak, 'The Strategic Corporal: Leadership in the Three Block War', *Marines Magazine*, January 1999.

Josh Kucera, 'Red Force Tracking Advances', *Jane's Defence Weekly*, 4 March 2005.

—— 'US Surveillance Link-up Will Boost Ability to Track the Enemy', *Jane's Defence Weekly*, 24 March 2005.

Philippe Lagassé, 'Tradition and Isolation: Canada, NorthCom and the UCP', p. 9, available at www.cda-cdai.ca/symposia/2002/lagasse.htm.

—— 'Northern Command and the Evolution of Canada-US Defence Relations', *Canadian Military Journal*, Spring 2003.

Darren Lake and Kim Burger, 'UK, USA Must do More to Reduce Friendly Fire Risk', *Jane's Defence Weekly*, 22 January 2003.

Jaron Lanier, 'Digital Maoism: The Hazards of New Online Collectivism', *Edge: The Third Culture*, 30 May 2006, available at www.edge.org/3rd_culture/lanier06/lanier06_index.html.

Scott Lash, *Critique of Information* (London: Sage Publications, 2002).

Karl Lautenschlager, 'Technology and the Evolution of Naval Warfare', *International Security*, vol. 8, no. 2, 1983.

Christopher Layne, 'America as European Hegemon', *The National Interest*, Summer 2003.

Robert Leonhard, *The Art of Maneuver: Maneuver Warfare Theory and AirLand Battle* (Novato, CA: Presidio Press, 1991).

Commodore Eric Lerhe (CF) and CPO2 Doug McLeod (CF), 'Canadian Naval Task Groups in Op Apollo', *Maritime Tactical Warfare Bulletin*, 2003.

Robert E. Levin, *The Global Information Grid and Challenges Facing Its Implementation*, GAO 84–858 (Washington DC: Government Accounting Office, July 2004).

Joris Janssen Lok, 'Next Level Needed for NATO ACCS', *International Defence Review*, 1 July 2002.

—— 'Communication Weaknesses Endanger Allied Integration in US led Air Campaigns', *International Defence Review*, 1 March 2004.

Douglas A. MacGregor, *Breaking the Phalanx* (Westport, CN: Praeger, 1997).

Mark MacIntyre and Sherri Flemming, 'Netcentric Warfare for Dynamic Coalitions: Implications for Secure Interoperability', paper presented at the RTO IST Symposium on 'Information Management Challenges in Achieving Coalition Interoperability', Québec, 28–30 May 2001.

Thomas MacIntyre, 'CENTRIXS Improves Communication for RIMPAC 2004', www.news.navy.mil, Story Number NNS040707–28, 8 July 2004.

Captain Paul Maddison (CF), 'The Canadian Navy's Drive for Trust and Technology in Network Centric Coalitions: Riding Comfortably Alongside, or Losing Ground in a Stern Chase?', course paper, Canadian Forces College, 2004.

Kishore Mahbubani, 'The Impending Demise of the Postwar System', *Survival*, vol. 47, no. 4, Winter 2005–2006.

Thomas G. Mahnken, 'Beyond Blitzkreig: Allied Responses to Combined-Arms Armoured Warfare during World War II', in Emily O. Goldman and Leslie C. Eliason (eds), *The Diffusion of Military Technology and Ideas* (Stanford, CA: Stanford University Press, 2003).

Marina Malenic, 'Allies May Want Role in Missile Defense Command and Control', *Inside the Army*, 21 May 2007.

Mark C. Malham and Deborah Gabbard, 'Battle Command Systems: The Force XXI Warfighter's Advantage', *Military Review*, vol. 78, no. 2, March/April 1998.

Michelle Malkin, 'The Photo Op Shop of Horrors', *The Washington Times*, 19 August 2006, www.washingtontimes.com/news/2006/aug/18/20060818–091848–7126r/.

Susan Maret, *On Their Own Terms: A Lexicon with an Emphasis on Information-Related Terms Produced by the U.S. Federal Government*, www.fas.org/sgp/library/maret.pdf.

Dwight Mason, 'Managing North American Defence at Home', paper presented at *What Canadian Military and Security Forces in the Future World? A Maritime Perspective*, Dalhousie University, 10–12 June 2005.

—— 'Time to Expand NORAD', *Security and Sovereignty: Renewing NORAD, One Issue, Two Voices #3*, Woodrow Wilson International Center for Scholars, 2005.

Susan C. McGovern, *Information Security Requirements for a Coalition Wide Area Network*, Masters thesis, Naval Post-Graduate School, Montery, June 2001.

Gary McKerow, 'Multilevel Security Networks: An Explanation of the Problem', *SANS Information Security Reading Room*, 5 February 2001.

Lt. Col. H. R. McMaster, 'Crack in the Foundation: Defense Transformation and the Underlying Assumption of Dominant Knowledge in Future War', *CSL Student Issue Paper*, Vol. S03–03, November 2003.

Walter Russell Mead, *Power Terror and War: American Grand Strategy in a World at Risk* (New York: Alfred A. Knopf, 2004).

Steven Metz, 'The Effect of Technological Asymmetry on Coalition Operations', in Thomas J. Marshall, Phillip Kaiser, and Jon Kessmeier (eds), *Problems and Solutions in Future Coalition Operations* (Carlisle, PA: US Army War College Strategic Studies Institute, December 1997).

James M. Minifie, *Peacemaker or Powdermonkey: Canada's Role in a Revolutionary World* (Toronto: McClelland & Stewart, 1960).

Ministry of Defence, *Network Enabled Capability* (London: HMSO, 2002).

Paul T. Mitchell, 'Small Navies and NCW: Is There a Role?', *Naval War College Review*, vol. 61, no. 2, Spring 2003.

—— 'International Anarchy and Coalition Interoperability in High Tech Environments', in David Carment and Martin Rudner (eds), *Peacekeeping Intelligence: New Players, Extended Boundaries* (Abingdon: Routledge, 2005).

—— '1000 Ship Navies, Maritime Domain Awareness, and Networks: The Policy Nexus', *RUSI Defence* Systems, vol. 10, no. 1, June 2007.

W. H. Moore, 'The United Kingdom's View of US Army Transformation', in Conrad C. Crane (ed.), *Transforming Defense* (Carlisle PA: Strategic Studies Institute, 2001).

Hans Morgenthau, *Scientific Man vs. Power Politics* (Chicago, IL: University of Chicago Press, 1946).

—— Alliances', in Julian R. Friedman, Christopher Bladen, and Steven Rosen (eds), *Alliance in International Politics* (Boston, MA: Allyn and Bacon Inc., 1970).

Jefferson Morris, 'UK Likely to Provide More Funding for DOD Sponsored NCW Study', *Aerospace Daily*, vol. 209, no. 12, 22 June 2004.

Jefferson Morris and Rich Tuttle, 'Contractors Lining Up To Compete for Transformational Communications Network', *Aerospace Daily*, vol. 207, no. 38.

Vincent Moscoe, *The Digital Sublime: Myth, Power, and Cyberspace* (Cambridge MA: MIT Press, 2004).

Multinational Interoperability Council, *Report of the Multinational Interoperability Council, 27–28 October 1999*, 1 March 2000.

—— *Report on MIC 2000, November 8–9, 2000*, 19 January 2001.

—— *Report on MIC 2002, April 16–18, 2002*, 7 June 2002.

Dennis Murphy, 'Networked Enabled Operations: Initial Impressions', *CSL Issue Papers*, Vol. 06–05, March 2005.

Williamson Murray, 'Armored Warfare: The British, French, and German Experiences', in Williamson Murray and Allan R. Millet (eds), *Military Innovation in the Interwar Period* (Cambridge: Cambridge University Press, 1996).

—— 'May 1940: Contingency and Fragility of the German RMA', in MacGregor Knox and Williamson Murray (eds), *The Dynamics of Military Revolution, 1300–2050* (Cambridge: Cambridge University Press, 2001).

Moisés Naim, 'The YouTube Effect', *Foreign Policy*, January/February 2007.

Bruce R. Nardulli *et al.*, *Disjointed War: Military Operations in Kosovo, 1999* (Santa Monica, CA: RAND Arroyo Center, 2002).

Greg Nash and David Stevens, *Australia's Navy in the Gulf* (Silverwater: Topmill, 2006).

National Audit Office, *Delivering Digital Tactical Communications through the Bowman CIP* (London: HMSO, 2006).

Nicholas Negroponte, *Being Digital* (New York: Vintage Books, 1995).

Lt. Col. John D. Nelson, 'Swiftly Defeat the Efforts: Then What?', *CSL Student Issue Paper*, Vol. S04–04, July 2004.

Jacob Neufeld, George M. Watson Jr, and David Chenoweth (eds), *Technology and the Air Force: A Retrospective Assessment* (Washington DC: USAF, 1997).

Scott L. Nicholas, 'Anti-carrier Warfare', in Bruce W. Watson and Susan M. Watson (eds), *The Soviet Navy: Strengths and Liabilities* (Boulder, CO: Westview, 1986).

Joseph Nye, 'Military De-Globalization', *Foreign Policy*, January–February 2001.

Office of Force Transformation, *US/UK Coalition Combat Operations during Operation Iraqi Freedom* (Washington DC: Department of Defense, 2005).

Robert E. Osgood, *The Entangling Alliance* (Chicago, IL: University of Chicago Press, 1962).

—— *Alliances and American Foreign Policy* (Baltimore: Johns Hopkins University Press, 1968).

Maj. John Owens, 'NEC: How Will it Affect the Philosophy of Mission Command?', student paper, Advanced Command and Staff Programme, Joint Services Command and Staff College.

Elias Oxendine IV, 'Managing Knowledge in the Battle Group Theatre Transition Process', student thesis, Naval Postgraduate School, Monterey, CA, September 2000.

Joe Pappalardo, 'Protecting GIG Requires a New Strategy', *National Defence*, October 2005.

Lieutenant Michael Parker (RAN), 'RAN Exercises', *Journal of the Australian Naval Institute*, no. 115, Summer 2005.

Zachery Petersen, 'NORAD Beginning to Develop Plan for New Maritime Warning Mission', *Inside the Pentagon*, 5 July 2006.

Lieutenant-Colonel William R. Pope, 'US and Coalition Command and Control Interoperability for the Future', thesis, US Army War College, Carlisle, PA, April 2001.

Barry R. Posen, 'The Battles of 1940', *The Sources of Military Doctrine* (Ithaca, NY: Cornell University Press, 1984).

—— 'Command of the Commons', *International Security*, vol. 28, no. 1, 2003.

David A. Powner and Eileen Laurence, *Information Sharing: The Federal Government Needs to Establish Policies and Processes for Sharing Terrorism-Related and Sensitive but Unclassified Information*, GAO-06–385 (Washington DC: GAO, March 2006).

David B. Ralston, *Importing the European Army* (Chicago, IL: University of Chicago Press, 1990).

Eric S. Raymond, 'The Cathedral and the Bazaar', www.catb.org/~esr/writings/cathedral-bazaar/cathedral-bazaar/.

Brock Read, 'Can Wikipedia Ever Make the Grade?', *Chronicle of Higher Education*, 27 October 2006, http://chronicle.com/tem/reprint.php?%20id=z6xht2rj60kqmsl8tlq5ltqcshc5y93y.

Ernie Regehr, *Canada and Ballistic Missile Defence* (Vancouver: Liu Institute for Global Issues, December 2003).

Major Joshua Reitz (USA), *Untangling the Web: Balancing Security, Prosperity, and Freedom in the Information Age*, MDS dissertation (Toronto: Canadian Forces College, May 2005).

Derek S. Reveron, 'Old Allies, New Friends: Intelligence Sharing in the War on Terror', *Orbis*, Summer 2006.

Howard Rheingold, *SmartMobs: The Next Social Revolution* (New York: Basic Books, 2002).

Tim Ripley, 'Part One: Communications – Speak Easy', *Jane's Defence Weekly*, 27 April 2005.

Karl Ritter, 'Report: Risk of Nuclear Warfare Rising', *Washington Post*, 11 June 2007, www.washingtonpost.com/wp-dyn/content/article/2007/06/11/AR2007061100663.html.

Robert W. Riscassi, 'Principles for Coalition Warfare', *Joint Forces Quarterly*, no. 1, Summer 1993.

John Robb, *Brave New War: The Next Stage of Terrorism and the end of Globalization* (Hoboken NJ: John Wiley and Sons, 2007).

Brian Robinson, 'DOD Arms Soldiers, Allies with Information', *Federal Computer Week*, vol. 20, no. 25, 31 July 2006.

Gene I. Rochlin, *Trapped in the Net: the Unanticipated Consequences of Computerization* (Princeton: Princeton University Press, 1997).

Bruce Rolfsen, 'Air Power Unleashed – Lessons from Iraq', *Armed Forces Journal*, 1 June 2003.

—— '191 F-15s Grounded at Least Another Month', *Air Force Times*, 11 January 2008.

Ann Roosevelt, 'Army Considers Improving Oneness of FBCB2 Systems', *C4I News*, 5 February 2004.

Emelie Rutherford, 'Officials Aim to Improve Global Maritime Situational Awareness', *Inside the Navy*, 16 April 2007.

Alan Ryan, 'Australian Army Cooperation with the Land Forces of the United States: Problems of a Junior Partner', *Land Warfare Studies Centre Working Paper*, no. 121, January 2003.

Claude Salhani, 'Cell Phone Cams Exposing Torture', *SpaceWar Daily*, 16 January 2007.

Greg Sandoval, 'Now Playing on the Net: War Propaganda', CNET News.com, 22 August 2006. http://news.cnet.com/Now-playing-on-the-Net-War-propaganda/2100–1038_3–6108004.html.

David Schmidtchen, *The Rise of the Strategic Private: Technology, Control and*

Change in a Network Enabled Military (Duntroon ACT: Land Warfare Studies Centre, 2006).

William B. Scott, 'Exercise Jump-Starts Response to Attacks', *Aviation Week and Space Technology*, 3 June 2002, available at www.aviationnow.com/content/publication/awst/20020603/avi_stor.htm.

Ann Scott Tyson, 'Military Goals Claim Priority over Diplomacy', *Christian Science Monitor*, vol. 93, no. 231, 24 October 2001.

Martin Shadwick, 'NORAD, Sovereignty, and Changing Technology', *YCISS Occasional Paper #3* (Toronto: York Centre for International and Strategic Studies, 1985).

Aarti Shah, 'McHale: Maritime NORAD Should be More Than Equivalence of Air Model', *Inside the Navy*, 13 June 2005.

John Shalikashvili, *Joint Vision 2010* (Washington DC: Joint Chiefs of Staff, 1997).

Scott Shane, 'Logged in and Sharing Gossip, er, Intelligence', *New York Times*, 2 September 2007, www.nytimes.com/2007/09/02/weekinreview/02shane.html.

William Shawcross, *Allies: The US, Britain, Europe and the War in Iraq* (London: Atlantic Books, 2003).

Jason Sherman, 'Bush Approves Updates to UCP, Assigns New Missions', *Inside the Army*, 5 June 2006.

Marvin Simpson and Leonard Simpson, 'Bettering National Response by Effectively Using the CAOC', paper presented at the Command and Control Research and Technology Symposium, 2006.

Maj. Jack L. Sine II (USA), *Organising the Fight: Technological Determinants of Coalition Command and Control and Combat Operations*, MA thesis, Naval Postgraduate School, September 2006.

M. Singer and A. Wildavsky, *The Real World Order: Zones of Peace/Zones of Turmoil* (Chatham, NJ: Chatham House, 1993).

Tony Skinner, 'UK Forces Deploy with Full Bowman Capability', *Jane's Defence Weekly*, 2 November 2005.

Edward Smith, 'Network Centric Warfare: What's the Point?', *Naval War College Review*, vol. 54, no. 1, Winter 2001.

Wing Commander H. Smythe, 'From Coningham to Project Coningham Keyes', student paper, Joint Command and Staff College, Watchfield UK.

Michael J. Sniffen, 'Controversial Government Data Mining Research Lives On', 23 February 2004, www.kdnuggets.com/news/2004/n05/20i.html.

Glenn H. Snyder, *Alliance Politics* (Ithaca, NY: Cornell University Press, 1997).

Lieutenant-Commander Thomas Spierto, 'Compromising the Principles of War: Technological Advancements Impact Multinational Military Operations', course paper, Naval War College, Newport, RI, 5 February 1999.

Sebastien Sprenger, 'US, Canadian Troops Could Respond Jointly to Terrorist Attacks', *Inside the Army*, 5 September 2005.

S. C. Spring *et al.*, 'Information Sharing for Dynamic Coalitions', unpublished paper, Pacific Sierra Research, Arlington, VA, December 2000.

Richard Stallman, 'Free Software: Freedom and Cooperation', 29 May 2001, www.gnu.org/events/rms-nyu-2001-transcript.html.

Bernard Stancati, 'The Future of Canada's Role in Hemispheric Defense', *Parameters*, Autumn 2006.

Nico Stehr, 'A World Made of Knowledge', available at www.inco.hu/inco0401/global/cikk1h.htm.

Statement by John P. Stenbit before the Committee on Armed Services, United States

House of Representatives, Terrorism, Unconventional Threats and Capabilities Subcommittee, 11 February 2004.

Robert B. Strassler, *The Landmark Thucydides* (New York: The Free Press, 1996).

Captain Robert M. Stuart (USN), 'Network Centric Warfare in Operation Allied Force: Future Promise or Future Peril?', course paper, Department of Joint Military Operations, US Naval War College, Newport, RI, 16 May 2000.

David Szabo and Todd M. Walters (eds), *The Canada US Partnership: Enhancing our Common Security. Workshop Report* (Washington DC: Institute for Foreign Policy Analysis, 2005), p. 9, available at www.ifpa.org/pdf/Canada-US-Report.pdf.

Don Tapscott and Anthony D. Williams, *Wikinomics: How Mass Collaboration Changes Everything* (New York: Penguin Group, 2006).

Curtis Taylor, 'Trading the Saber for Stealth – Can Surveillance Technology Replace Traditional Aggressive Reconnaissance?', *The Land Warfare Papers*, no. 53, September 2005.

Clive Thompson, 'Open Source Spying', *New York Times Magazine*, 3 December 2006, www.nytimes.com/2006/12/03/magazine/03intelligence.html.

Alvin and Heidi Toffler, *War and Anti-war: Survival at the Dawn of the 21st Century* (Boston, MA: Little, Brown, 1993).

Timothy Travers, *The Killing Ground: The British Army, the Western Front, and the Emergence of Modern Warfare, 1900–1918* (London: Allen Unwin, 1987).

Lisa Troshinsky, 'Force Networking in Operation Iraqi Freedom a Big Improvement, Official Says', *Aerospace Daily*, vol. 209, no. 1, 26 January 2004.

Gordon Trowbridge, 'Bringing Order from Chaos', *Air Force Times*, 20 December 2004.

Marita Turpin and Niek du Plooy, 'Decision-Making Biases and Information Systems', *Decision Support in an Uncertain and Complex World: The IFIP TC8/WG8.3 International Conference*, http://vishnu.sims.monash.edu.au:16080/dss2004/proceedings/pdf/77_Turpin_Plooy.pdf.

Mark Unewisse, Paul Gaertner, Anne Marie Grisogono and Robert Seymore. 'Land Situational Awareness for 2010', Defence Science and Technology Organisation paper, Australian Defence Force, p. 2, available at www.siaa.asn.au/get/2395379372.pdf.

Thomas Valovic, *Digital Mythologies: The Hidden Complexities of the Internet* (Piscataway NJ: Rutgers University Press, 2000).

Milan Vego, *Operational Warfare* (Newport, RI: Naval War College, 2000).

Stephen Walt, *The Origins of Alliances* (Ithaca: Cornell University Press, 1987).

Kenneth Waltz, 'Globalization and American Power', *The National Interest*, Spring 2000.

David Weinberger, *Loosely Joined Pieces: A Unified Theory of the Web* (Cambridge MA: Perseus Publications, 2002).

Norbert Weiner, *Cybernetics; Or, Control and Communication in the Animal and the Machine* (New York: Wiley, 1948).

James A. Winnefeld and Dana J. Johnson, *Joint Air Operations: Pursuit of Unity in Command and Control* (Annapolis: Naval Institute Press, 1993).

Captain Phil Wisecup and Lieutenant Tom Williams (USN), 'Enduring Freedom: Making Coalition Naval Warfare Work', *Proceedings*, vol. 128, no. 9, September 2002.

Paul Wolfowitz, 'Remembering the Future', *The National Interest*, Spring 2000.

—— 'Global Information Grid (GIG) Overarching Policy', *Department of Defense Directive 8100.1*, 19 September 2002, www.dtic.mil/whs/directives/corres/html2/d81001x.htm.

Michele Zanini and Jennifer Morrison Taw, *The Army and Multinational Force Compatibility* (Santa Monica: Rand Arroyo Center, 2000).

Rear-Admiral Thomas E. Zelibor (USN), 'FORCEnet is Navy's Future: Information Sharing from Seabed to Space', *Armed Forces Journal*, December 2003, www.chinfo.navy.mil/navpalib/.www.rhumblines/rhumblines170.doc.

Jonathon Zittrain, *The Future of the Internet – And How to Stop It* (New Haven CT: Yale University Press, 2008).

Index